CORD ALGEBRA 1

PART B 1

Mathematics in Context

**Developed by the
Center for Occupational Research and Development
Waco, Texas**

JOIN US ON THE INTERNET
WWW: http://www.thomson.com
EMAIL: findit@kiosk.thomson.com   A service of I(T)P®

South-Western Educational Publishing
an International Thomson Publishing company I(T)P®

Cincinnati • Albany, NY • Belmont, CA • Bonn • Boston • Detroit • Johannesburg • London • Madrid
Melbourne • Mexico City • New York • Paris • Singapore • Tokyo • Toronto • Washington

CORD Staff
Director: John Souders, Jr., Ph.D.
Mathematics Specialist: Michael Crawford, Ph.D.
Educational Specialist: Lewis Westbrook
Chief Scientist: Leno Pedrotti, Ph.D.
Applications Specialist: John Chamberlain

South-Western Educational Publishing Staff
Publisher/Team Leader: Thomas Emrick
Project Manager: Suzanne Knapic
Marketing Manager: Colleen Skola
Consulting/Developmental Editor: Richard Monnard
Manufacturing Coordinator: Jennifer Carles

Consultants

Wes Evans
Mathematics Teacher
Midway High School
Waco, Texas

Donna Long
Mathematics Coordinator
MSD Wayne Township
Indianapolis, Indiana

Editorial Services: Monotype Editorial Services
Production Services: Monotype Composition Company, Inc.
Design: Zender + Associates, Inc.

Copyright © 1998
by the Center for Occupational Research and Development
Waco, Texas

Published and distributed
by South-Western Educational Publishing
5101 Madison Road
Cincinnati, OH 45227
1-800-354-9706 Fax: 1-800-487-8488

ISBN: 0-538-67466-0

2 3 4 5 6 7 8 WS 01 00 99 98

Printed in the United States of America

I(T)P®

International Thomson Publishing Company
South-Western Educational Publishing is an ITP Company. The ITP trademark is used under license.

CHAPTER 7

Statistics and Probability

Featuring

CHAPTER 8

Systems of Equations

APPLICATIONS AND CONNECTIONS

CHAPTER 9

Inequalities

CHAPTER 10

Polynomials and Factors

CHAPTER 11

Quadratic Functions

CHAPTER 12

Right Triangle Relationships

SKILLS REVIEW

Congratulations. You have finished the first half of *CORD Algebra 1*. Before you begin Part B, stop and review some of the important skills presented in Part A. First, read the Skill Review; then, complete the exercises. This will give you a good foundation for a successful journey through the second half of *CORD Algebra 1*.

Skill 1 Adding and Subtracting Integers

When two integers have the same sign, add their absolute values. Use the common sign.

$$5 + 8 = 13 \qquad\qquad (-6) + (-7) = -13$$

When two integers have opposite signs, find the difference of their absolute values. Use the sign of the integer with the greater absolute value.

$$9 + (-4) = 5 \qquad\qquad -8 + 3 = -5$$

To subtract an integer, add its opposite.

$$12 - (-4) = 12 + 4 = 16 \qquad\qquad -7 - (-6) = -7 + 6 = -1$$
$$15 - 3 = 15 + (-3) = 12 \qquad\qquad -9 - 6 = -9 + (-6) = -15$$

Perform the indicated operations from left to right.

1. $-12 - 9$ **2.** $15 - 22$ **3.** $-14 + (-17)$

4. $3 + 8 - 6$ **5.** $-6 + 4 - (-3)$ **6.** $-5 - 7 - (-6)$

7. $23 - 48 + 35$ **8.** $6.1 - 4.8 - 4.2$ **9.** $-\$5.60 + \$3.33 - \$8.65$

Skill 2 Multiplying and Dividing Integers

When two integers have the same sign, their product or quotient is positive.

$$8 \times 9 = 72 \qquad\qquad -6 \times -7 = 42$$
$$36 \div 12 = 3 \qquad\qquad -48 \div -6 = 8$$

When two integers have opposite signs, their product or quotient is negative.

$$7 \times (-3) = -21 \qquad\qquad -15 \times 4 = -60$$
$$35 \div -5 = -7 \qquad\qquad -50 \div 2 = -25$$

Perform the indicated operations from left to right.

1. 17×-4 **2.** $-108 \div -6$ **3.** $-6.2 \div -3.1$

4. -2.5×-1.2 **5.** $39 \div -1.3$ **6.** -0.5×-3.24

7. $-8 \times -3 \times -2$ **8.** $-45 \div 9 \times -6$ **9.** $-100 \div -5 \div -5$

SKILLS REVIEW

Skill 3 Writing Large Numbers in Scientific Notation

To convert a large number from decimal notation to scientific notation:

1. Start from the left. Place a pointer to the right of the first nonzero digit.

$$16,500,000.$$
$$\uparrow$$

2. Count the spaces as you move from the counter to the decimal point.

$$16,500,000.$$
$$\uparrow \ 7 \text{ spaces}$$

3. Move the decimal point to the pointer and multiply the number by the power of ten equal to the spaces you moved.

$$1.65 \times 10^7$$

Convert each amount to scientific notation.

1. 6,200,000   **2.** 425,000   **3.** 360,000,000

4. 19,256,000   **5.** 158,600,000   **6.** 8,000,000,000

7. The average distance from the Earth to the sun is thirty-three million miles.

8. The national debt is over four trillion dollars.

Skill 4 Writing Small Numbers in Scientific Notation

To convert a small number from decimal notation to scientific notation:

1. Start from the left. Place a pointer to the right of the first nonzero digit.

$$0.00035$$
$$\uparrow$$

2. Count the spaces as you move from the decimal point to the pointer.

$$0.00035$$
$$\uparrow \ 4 \text{ spaces}$$

3. Move the decimal point to the pointer and multiply by the power of $\frac{1}{10}$ equal to the spaces you moved.

$$3.5 \times 10^{-4}$$

Convert each amount to scientific notation.

1. 0.016   **2.** 0.0004   **3.** 0.000007

4. 0.000315   **5.** 0.0000009   **6.** 0.000024

SKILLS REVIEW

Skill 5 Evaluating Expressions

To evaluate an expression, substitute the given value for the variable and perform the operations according to the order of operations rule.

- First, perform all operations within parentheses.
- Second, perform all operations involving exponents.
- Third, multiply and divide in order from left to right.
- Finally, add and subtract in order from left to right.

$$-3 \times (4 \div 2)^2 + 1 = -3 \times 2^2 + 1 = -3 \times 4 + 1 = -12 + 1 = -11$$

Evaluate each expression.

1. $10 \div (y + 1)$ when y is -6

2. $-5(b \div 3)$ when b is -6

3. $4 + 3(m - 1)$ when m is -2

4. $2 + (t + 4)^2 \div 5$ when t is 1

5. $5(3a + 2)^2 + 9$ when a is -3

6. $8 + -4(c - 5)^3 \div 2$ when c is 3

Skill 6 Using Formulas

A formula is a general rule expressed with letters written in the form of an equation.

The formula for the circumference of a circle:	$C = 2\pi r$
The formula for the area of a circle:	$A = \pi r^2$
The formula for the volume of a rectangular box:	$V = LWH$
The formula for the volume of a sphere:	$V = \frac{4}{3}\pi r^3$
The formula for simple interest:	$I = PRT$

Find

1. The circumference of a circle if the radius is 2.4 meters.

2. The area of a circle if the radius is 3 inches.

3. The volume of a rectangular box if the length is 2 feet, the width is 3 feet, and the height is 5 feet.

4. The volume of a sphere if the radius is 5 centimeters.

5. The simple interest if the principal is $500, the rate is 9.5%, and the time is 3 years.

Skill 7 Solving Multiplication and Division Equations

If each side of an equation is multiplied by the same number or divided by the same nonzero number, the results are equal; that is, the equation is still balanced.

$$\text{Solve } 3b = 9 \qquad\qquad \text{Solve } \frac{c}{-8} = 2$$

$$\frac{3b}{3} = \frac{9}{3} \qquad\qquad -8 \cdot \frac{c}{-8} = -8 \cdot 2$$

$$b = 3 \qquad\qquad c = -16$$

Solve each equation.

1. $5q = -30$ **2.** $\frac{k}{4} = -3$ **3.** $\frac{2}{3}m = 12$

4. $\frac{d}{7} = 1$ **5.** $1\frac{1}{2}r = -6$ **6.** $-8p = -40$

7. Solve $I = PRT$ for T.

8. Solve $\frac{d}{r} = t$ for d.

Skill 8 Solving Addition and Subtraction Equations

If the same number is added to or subtracted from each side of an equation, the results are equal; that is, the equation is still balanced.

$$\text{Solve } a - 6 = 9 \qquad\qquad \text{Solve } b + 4 = -6$$

$$a - 6 + 6 = 9 + 6 \qquad\qquad b + 4 - 4 = -6 - 4$$

$$a = 15 \qquad\qquad b = -10$$

Solve each equation.

1. $24 + q = 35$ **2.** $-15 + v = -25$ **3.** $9 = s + 15$

4. $p - 4.4 = 8.3$ **5.** $t - 8.1 = -5.3$ **6.** $-p + 3.4 = 1.7$

7. Solve $a + b = c$ for a.

8. Solve $a - b = c$ for a.

SKILLS REVIEW

Skill 9 Solving Multi-Step Equations

Many equations require more than one operation to solve. Isolate the variable and simplify the constants on the other side of the equal side. Use the order of operations.

$$\text{Solve } 5n + 4 = 9 \qquad\qquad \text{Solve } -3(n - 4) = -6$$

$$5n + 4 - 4 = 9 - 4 \qquad\qquad -3n + 12 = -6$$

$$5n = 5 \qquad\qquad -3n + 12 - 12 = -6 - 12$$

$$\frac{5n}{5} = \frac{5}{5} \qquad\qquad -3n = -18$$

$$n = 1 \qquad\qquad n = 6$$

Solve each equation.

1. $3m - 4 = 11$ **2.** $6(b + 2) = 12$ **3.** $1\frac{1}{2}x + 5 = 179$

4. $-24 = -3(2m + 1)$ **5.** $\frac{3}{4}(x + 2) = -6$ **6.** $\frac{8c + 2}{2} = -15$

Skill 10 Solving Equations with Variables on Both Sides

The Distributive Property can be used to add and subtract like terms.

$$5a + 6a = (5 + 6)a = 11a \qquad\qquad 8b - 3b = (8 - 3)b = 5b$$

To solve an equation with variables on both sides, isolate the variables on the same side. Combine like terms and solve.

$$\text{Solve } 6m = 2m + 4$$

$$6m - 2m = 2m - 2m + 4$$

$$4m = 4$$

$$m = 1$$

Solve each equation.

1. $5t + 2t = -21$ **2.** $-6r = 3r + 18$ **3.** $12 - 2y = 4y$

4. $3(n + 2) = 2n$ **5.** $b - 4 = 3(b + 2)$ **6.** $4(b - 5) = 2(b + 3)$

Skill 11 Locating Points on a Coordinate System

An ordered pair of numbers (x,y) called coordinates are used to locate points on a coordinate system. The x-value gives the distance to the left or right of the origin. The y-value gives the distance above or below the origin. The point A (2,3) is 2 units to the right of the origin and 3 units above the origin.

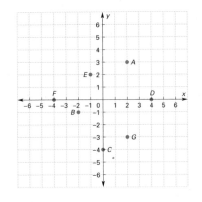

Name the coordinates of each point.

1. B **2.** C **3.** D

4. E **5.** F **6.** G

7. Explain how to locate S $(-2,4)$.   **8.** Explain how to locate T $(-5,-5)$.

Skill 12 Finding the Slope of a Line

The slope of a line is the ratio of its rise to its run. If A (a,b) and B (c,d) are two points on a line, the slope of the line is

$$\frac{b - d}{a - c} \quad \text{or} \quad \frac{d - b}{c - a}$$

The slope of line AB is

$$\frac{3 - 1}{4 - 2} = \frac{2}{2} = 1$$

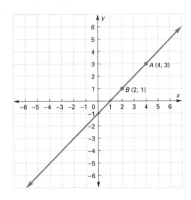

Find the slope given the rise and the run.

1. rise 4; run -2 **2.** rise -1; run -5 **3.** rise 3, run -12

Find the slope of the line that passes through each pair of points.

4. A (5,3) and B (2,6)   **5.** C $(-4,1)$ and D $(2,-6)$ **6.** E $(-3,-5)$ and F $(0,-3)$

7. Explain how to find the slope of the line passing through G $(-s,t)$ and H $(u,-v)$.

8. The slope of a line is $\frac{2}{3}$. One point on the line is X (3,5). List two more points on the line.

SKILLS REVIEW

Skill 13 Finding the Intercepts of a Line

The graph of an equation written in slope-intercept form $y = mx + b$ is a line with slope m that intersects the y-axis at the point $(0,b)$. The value of b is called the y-intercept of the line. To find the slope and y-intercept for $4x - 5y = -20$, write the equation in slope intercept form

$$4x - 5y = -20$$

and solve for y:

$$y = \frac{4}{5}x + 4$$

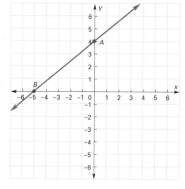

The slope is $\frac{4}{5}$, and the y-intercept is 4. To find the x-intercept, substitute 0 for y and solve for x. If $4x - 5(0) = -20$, $x = -5$. The x-intercept is -5. On the graph, A is $(0,4)$ and B is $(-5,0)$.

Find the slope and y-intercept of each line.

1. $y = 5x + 8$ **2.** $2x + y = 6$ **3.** $3x = 5y - 10$

Find the x- and y-intercept of each line.

4. $3x - 4y = 12$ **5.** $2y + 3x = 10$ **6.** $x - 8y = -4$

Skill 14 Finding the Equation of a Line

The equation of a line can be determined if you know any two points on the line. First, find the slope of the line through A and B.

$$m = \frac{8 - 4}{3 - 1} = \frac{4}{2} = 2$$

Let $P(x,y)$ be another point on the line passing through A and B. Find the slope determined by P and either of the other two points.

$$m = \frac{y - 4}{x - 1}$$

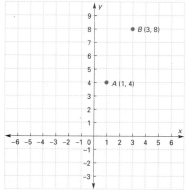

Thus, $\frac{y - 4}{x - 1} = 2$, and the equation of the line is $y - 4 = 2x - 2$, or $y = 2x + 2$.

Find the equation of the line through the given points.

1. $(1,2)$ and $(2,4)$ **2.** $(0,4)$ and $(2,6)$ **3.** $(1,4)$ and $(6,12)$

4. $(-2,0)$ and $(3,-4)$ **5.** $(-1,3)$ and $(2,2)$ **6.** $(-3,-4)$ and $(3,4)$

SKILLS REVIEW

Skill 15 Graphing the Absolute Value Function

The absolute value of a number is its distance from zero on a number line. Thus,

$$\text{If } x \geq 0, |x| = x$$

$$\text{If } x < 0, |x| = -x$$

The absolute value function is written $y = |x|$ or $f(x) = |x|$. The graph of the parent absolute value function looks like a V with its vertex at V(0,0).

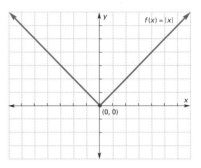

Describe the changes in the graph of the parent function for each equation.

1. $y = |x| + 1$ **2.** $y = |x + 1|$ **3.** $y = |x| - 3$

4. $y = 2|x|$ **5.** $y = -3|x|$ **6.** $y = |x - 1| + 2$

Skill 16 Graphing Quadratic Functions

The parent quadratic function is written

$$y = x^2 \text{ or } f(x) = x^2$$

The graph of $y = x^2$ is a parabola with its vertex at V(0,0).

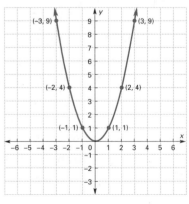

Describe the changes in the graph of the parent function for each equation.

1. $y = x^2 + 1$ **2.** $y = (x + 1)^2$ **3.** $y = x^2 - 3$

4. $y = 2x^2$ **5.** $y = -3x^2$ **6.** $y = (x - 1)^2 + 2$

CORD ALGEBRA 1

PART AB 1

Mathematics in Context

CHAPTER 7

WHY SHOULD I LEARN THIS?

Information will continue to grow as technology expands. In this chapter, you will explore how the principles of statistics and probability are used to work with large amounts of data. Learn how statistics and probability can help you manage information and expand your career options.

STATISTICS AND PROBABILITY

OBJECTIVES

1. Calculate the mean, median, and mode for a set of data.
2. Draw a histogram from a frequency distribution table.
3. Find the probability of a single event.
4. Find the probability of mutually exclusive events.
5. Find the probability of dependent and independent events.

Computer networking and the Internet have saturated the world with numerical data. People must have quick ways to analyze that data and extract the information it contains. **Statistics** is the branch of mathematics that helps organize and find meaning in numerical data.

For example, teachers collect a large amount of numerical data each time a test is given. They use statistics to analyze this data and provide insights into how individual students are performing with respect to the rest of the class.

Probability is useful in predicting outcomes of a future event. For example, on election night, news reporters often make predictions on which of the candidates will win an office. To make these predictions, telephone surveys are conducted to collect data on how small, specially selected groups of people voted. These voter responses are analyzed and compared with past voting patterns to determine the probability of a certain candidate winning. From these probabilities, political analysts forecast winners in elections.

As you watch the video for this chapter, notice how people on the job use statistics and probability to solve problems.

LESSON 7.1 MEASURES OF CENTRAL TENDENCY

You are the production supervisor in a small manufacturing plant that produces handmade skateboards. To measure the productivity of three workers, you make a table to record how many skateboards each worker completes in 10 successive work days.

Production Data

Worker	\ Day of the Week M	T	W	Th	F	M	T	W	Th	F
A	1	2	2	3	3	4	5	4	5	5
B	6	1	2	5	3	2	3	2	7	1
C	7	6	5	4	2	3	2	3	2	2

Now that you have organized the data in a table, what are some of the typical values you can find that give meaning to the data? For example, for each worker,

a. Which number occurs most often?

b. If a worker's production numbers are arranged in order, what is the middle number?

c. What is the average number of skateboards produced?

The answer to each of these questions is a statistic called a **measure of central tendency**.

> **Measures of Central Tendency**
> A measure of central tendency is a number used to represent all the numbers in a data set.

The mode, the median and the mean are the most commonly used measures of central tendency. Use the questions **a**, **b**, and **c** above to write a definition of mode, mean, and median. Give some examples where you would use each measure during a typical day.

The Mode

Look at the data in the table for Worker A. Which number occurs most often? You can see from the data set that 5 occurs three times. The other numbers occur two times or less.

The number that occurs most frequently in a set of numbers is called the **mode** of the set. The mode is 5 skateboards per day for Worker A.

> **Mode**
> The mode is the number that occurs most often in a set of numbers.

Ongoing Assessment

Use the data table to find the mode for Workers B and C.

EXAMPLE 1 Zero Mode or Two Modes

Find the mode for data set **a** and **b**.

a.   5   6.2   7   8   9.4

b.   4   4   5   6   6
         6   7   8   8   8
         9   10

SOLUTION

Data set **a** does not have a mode. No one number in this set occurs more frequently than any of the other numbers.

Data set **b** has two modes, 6 and 8. Both 6 and 8 occur most frequently (three times).

A set of numbers may have no mode, one mode, or more than one mode. You can usually find the mode by quickly looking at the data.

The mode is meaningful when there is a large number of values in the data set. The mode represents the *most typical* value in a large set of data.

Suppose you are responsible for inventory control at an office supply store. You may be very interested in what kind of paper people are using the most; that is, you might want to know the modes for paper weights and sizes.

The Median

Another measure of central tendency is the **median**.

To find the median of the data for Worker A, first list the data in order from greatest to least.

5   5   5   4   4   3   3   2   2   1

The median divides the data into two equal groups. In other words, there are as many numbers above the median point as there are below it. Since there are ten numbers in all (for ten days), look for the point that has five numbers above it and five numbers below it. Draw a line to show where this happens.

5 5 5 4 4 | 3 3 2 2 1

Where does the line fall? It falls halfway between 3 and 4. Therefore, the median production for Worker A is 3.5 skateboards.

Notice that the median represents the midpoint on a scale of measurement. The median does not necessarily equal an actual data value. Thus, the median for Worker A's daily production is 3.5 skateboards. Yet, Worker A never actually produced 3.5 skateboards on any of the ten days.

Median

The median is the middle number of a data set, when the data are arranged in numerical order. If there is no middle number, the median is the average of the two middle numbers.

Ongoing Assessment

Use the data table to find the median for Workers B and C.

The median is a useful way to show the central tendency for data that may have some extreme values. These values are not really typical of the rest of the data. These extreme values are called **outliers**. An outlier can result from unusual circumstances during the collection of data. For example, you are collecting data on the heights of students in your class, and one of your class members is the center of the basketball team. This height may be an outlier in your data set.

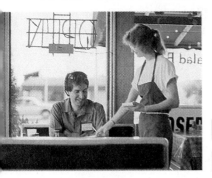

A person serving tables in a restaurant recorded these tips on 7 different days.

Day 1	Day 2	Day 3	Day 4	Day 5	Day 6	Day 7
$24	$15	$22	$80	$16	$21	$19

1 What is the mode for the data?

2 What is the median for the data?

Critical Thinking   What are outliers in the data set in Activity 1. What effect does an outlier such as $80 have on using the median to report the daily tips received?

The Mean

A third measure of central tendency (the one used most often) is called the **mean**, or average.

> **Mean**
> The mean or average of a set of data is the sum of all the values of the data divided by the number of elements in the data set.

EXAMPLE 2 Finding a Mean

Find the mean of the tips from Activity 1.

Day 1	Day 2	Day 3	Day 4	Day 5	Day 6	Day 7
$24	$15	$22	$80	$16	$21	$19

SOLUTION

First, add all the tips. Use your calculator to help. The total is $197.

Second, divide the total ($197) by the number of days (7)

The mean value for the data is about $28.14 per day.

You can now find three measures of central tendency for a set of data. Each measure gives useful information about the data.

Critical Thinking A friend is thinking about taking a job at the same restaurant and asks, "How much in tips do you usually get for a day?" For the data set given in Example 2, which measure of central tendency would you use to answer this question? Explain your answer.

ACTIVITY 2 Comparing Measures of Central Tendency

1 Complete the table using the numbers from the production data table.

Three Measures of Central Tendency			
	Mode	**Median**	**Mean**
Worker A	5	3.5	?
Worker B	?	?	?
Worker C	?	?	?

2 Circle the largest number in each column.

3 Which worker(s) has the highest mode?

4 Which worker(s) has the highest median?

5 Which worker(s) has the highest mean?

6 Suppose you must move two workers to another job. Which worker would you not move? Explain your answer.

Critical Thinking Here are three examples where the mode, median or mean can be used to describe a situation. Which measure of central tendency would you use for each one, and why?

1. A supermarket manager wants to know how many boxes of cereal customers purchase most often.

2. A real estate agent wants to set the selling price of a three-bedroom house so that there are as many comparable houses above as below the price.

3. Students want to know the central tendency that tells the average of all their test scores in a certain grading period.

LESSON ASSESSMENT

Think and Discuss

1 Explain what is meant by "measures of central tendency."

2 Describe a situation in which the mode is the best measure of central tendency.

3 Describe a situation in which the median is the best measure of central tendency.

4 Describe a situation in which the mean is the best measure of central tendency.

5 Create a set of data using five numbers in which the mean, the median, and the mode are the same number.

Practice and Problem Solving

Mary's first semester pop quiz scores are:

5, 7, 7, 9, 10, 10, 10, 12, 14

Find the

6. Mode.   **7.** Median.   **8.** Mean.

Mary's test scores for the semester are:

85, 75, 70, 95, 55, 60, 100, 70, 85, 70, 65, 70, 85

Find the

9. Mode.   **10.** Median.   **11.** Mean.

The semester grade point averages for Mary and her friends are:

2.9, 2.3, 2.5, 2.4, 2.9, 2.7, 2.4, 2.4, 2.3, 2.4

Find the

12. Mode.   **13.** Median.   **14.** Mean.

Shelly Lynn records the number of cellular phone subscriptions she sells in different cities:

355, 262, 285, 237, 262, 258, 275

Find the

15. Mode.   **16.** Median.   **17.** Mean.

At the beginning of an aerobics class, the instructor recorded the weights of the participants.

Tom 58 kg, Joe 60 kg, Ann 60 kg, Jaren 49 kg, Nancy 63 kg, Kim 60 kg, Chris 67 kg, Megan 58 kg

18. Record these names and weights in a table from the least weight to the greatest weight.

19. Find the mode for this data.

20. Find the median for this data.

21. Find the mean or average weight for the class.

22. Suppose John, who weighs 65 kg, joins the class. Without computing, estimate how John's weight will affect the mean you found in Exercise 21.

23. Will John's weight affect the mode or median? Explain why or why not.

Mixed Review

24. The speed of light is approximately 299,000,000 meters per second. Write this number in scientific notation.

25. Laura is going to loan her brother $5000 to use as a down payment on a new car. If her brother borrows the money at 8% simple interest, how much money will he owe Laura at the end of 6 months?

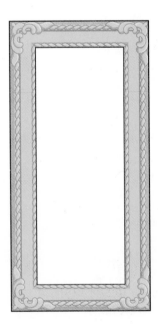

26. The formula $C = \dfrac{A}{46}$ gives the number of circuits required to wire a room of area A. Solve the formula for A.

Write and solve an equation for each situation.
27. After Thomas wrote a check for $249, he had $878.45 in his checking account. How much did Thomas have in his account before he wrote the check?

28. The length of a rectangular picture frame is twice its width. The perimeter of the frame is 108 inches. Find the length and width of the picture frame.

LESSON 7.2 FREQUENCY DISTRIBUTIONS

In Lesson 7.1, you used a data table to organize production data for three workers.

					Skateboard Production Data					
				Day of the Week						
Worker	M	T	W	Th	F	M	T	W	Th	F
A	1	2	2	3	3	4	5	4	5	5
B	6	1	2	5	3	2	3	2	7	1
C	7	6	5	4	2	3	2	3	2	2

Organizing Data

A **frequency distribution** is another useful tool for organizing information. To make a frequency distribution table for Worker A, start by separating the data into five classes. Each class represents a different production level. For example, the class labeled 5 means a worker produces 5 skateboards in a day.

Next, tally (count) the number of times Worker A meets the production levels specified by the classes. For example, on the first Monday Worker A produced 1 skateboard. Therefore, place a tally mark in the class labeled 1. Continue through the rest of the day's list, placing tally marks in the appropriate classes. Finally, determine the class frequency by counting the number of tallies in each class. These steps will produce the following frequency distribution table.

Production Data for Worker A		
Classes	Tally	Frequency
5	///	3
4	//	2
3	//	2
2	//	2
1	/	1

To find the mean from a frequency distribution table, first multiply the class number by the frequency. If you let c represent the class and f represent the frequency, this product is

$$c \bullet f$$

Next, add each of the products. The Greek symbol Σ (sigma) means to add a set of numbers. Thus, the second step is to find

$$\Sigma(c \bullet f)$$

Finally, divide the sum by the total number n of tallies. The result is the mean and is usually denoted by $\bar{x}$. The formula for finding the mean of a frequency distribution is written as follows.

Formula for the Mean

Let c represent the class, f the frequency, and n the number of tallies. To calculate the mean $\bar{x}$ from a frequency distribution table, use this formula:

$$\bar{x} = \frac{\Sigma(c \bullet f)}{n}$$

EXAMPLE 1 Finding the Mean

Use the formula to find the mean of Worker A's production.

SOLUTION

Complete the table.

Classes	Tally	Frequency	$(c \bullet f)$
5	///	3	15
4	//	2	8
3	//	2	6
2	//	2	4
1	/	1	1
		10	34

$$n = 10 \qquad \Sigma(c \bullet f) = 34$$

Thus, $\bar{x} = \dfrac{34}{10} = 3.4$.

Critical Thinking How can you find the mode and the median by using the tally marks?

Ongoing Assessment

Make a frequency distribution table for Worker B and Worker C. Use the formula to find the mean for each worker. Use the tally marks to find the mode and median for each worker.

EXAMPLE 2 Using Data from a Frequency Table

Use the frequency distribution table in Example 1.

a. On how many days did Worker A make fewer than three skateboards?

b. On what percent of the days did Worker A make more than the mean of the distribution?

SOLUTION

Examine the frequency distribution table in the columns labeled "classes" and "frequency." The number in the frequency column is the number of days the worker equaled the production level in the classes column.

a. Worker A made 1 skateboard on one day and 2 skateboards on two days. Thus, Worker A made fewer than 3 skateboards three days.

b. The mean is 3.4. Worker A made 4 skateboards on two days and 5 skateboards on three days. Thus, Worker A made more than the mean on five out of ten days, or 50% of the days.

Histograms

A graph of the frequency distribution is a very useful means of displaying large sets of data. Consider the weights (in pounds) of thirty football players.

286, 234, 211, 268, 227,
273, 276, 250, 184, 230,
228, 202, 260, 205, 193,
250, 248, 234, 234, 197,
246, 224, 218, 235, 235,
253, 219, 241, 226, 246

1 Identify the largest and smallest numbers in the group, 286 and 184. The difference between these numbers is the **range** of the data. What is the range for this data?

2 Group the data into weight intervals that each contain ten pounds. This is how the table begins:

Weight	Tally	Frequency
280–289	/	1
270–279	//	2

3 Complete the table on your paper.

4 Draw a positive x-axis and a positive y-axis.

5 Label the x-axis as "Weights of Players". The values on the x-axis begin with the weight interval 180–189 and end with 280–289.

6 Label the y-axis as "Frequencies." The frequencies begin with zero and go up to 6.

7 For each weight interval, draw a rectangle to show the frequency. Since each weight interval has equal value, the widths of the rectangles are the same.

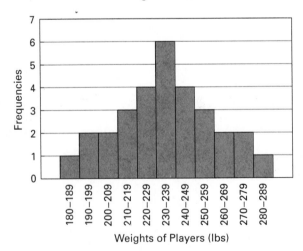

8 The height of each rectangle represents the frequency. The tallest rectangle represents the weight interval that occurs most often. Which weight interval is the mode for the weights of the football players?

Scientific or graphics calculators can evaluate many of the formulas used in statistics. Use your calculator to find the mean in the Activity.

The graphics calculator uses a menu key with the abbreviation VARS. You can use the variable statistics key to find the mean ($\bar{x}$), the sum (Σ), and the number of data values (n).

WORKPLACE COMMUNICATION

For the July 15 data, which time interval represents the mode for bandwidth usage? If the rate of increase in the use of the Internet continues, how long do you think it will be until MWI exceeds the capacity (100% of available bandwidth) of its T-1 connection? Why is there a "dip" in the middle of the histograms?

ALERT   ALERT   ALERT   ALERT   ALERT

THIS IS AN AUTOMATIC ALERT MESSAGE FROM

ServPro

YOUR INTERNET SERVICE PROVIDER

Modern Workplace, Inc.
Attention: Computer Network System Administrator

Dear System Administrator:

This automatic alert message has been generated because your system has exceeded 70% of the capacity of your T-1 connection to *ServPro* and the Internet.

We detected your system above 70% of available bandwidth on July 15, during the hours of operation from 7AM to 7PM. Data for July 15, as well as six months prior, are shown below.

The number of Internet users on the MWI computer network has increased significantly over the past six months. If the increase continues, we highly recommend you install another T-1 connection to *ServPro*. We are standing by to assist you in this matter. Please call your *ServPro* technical representitive, Ms. Janet Porter, at 1-800-*SERVPRO* for further advice and assistance.

LESSON ASSESSMENT

Think and Discuss

1 What is a frequency distribution?

2 How is the expression $\Sigma (c \cdot f)$ used?

3 Explain how to find the mean of a frequency distribution.

4 Explain how to find the mode of a frequency distribution from a histogram.

Practice and Problem Solving

In one month a landscape company trimmed trees on 20 different days. The supervisor recorded the number of trees trimmed each day.

	M	Tu	W	Th	F
Week 1	10	30	7	5	25
Week 2	24	28	11	17	16
Week 3	19	28	21	26	35
Week 4	3	4	21	13	9

5. Use the data to make a frequency distribution table.

6. Make a histogram of the data.

For the data, find the

7. Range.

8. Mode.

9. Median.

10. Use the formula to find the mean of the data.

The Johnson Company manufactures linen goods. In one week, the company packed and shipped the following number of linen packages to 25 retail customers.

80, 95, 50, 85, 100, 50, 75, 75, 80, 85, 70, 90, 95, 80, 85, 75, 80, 85, 70, 80, 100, 85, 60, 80, 95

11. Set up a frequency table and tally the number of packages.

12. Total the frequencies.

13. Make a histogram.

14. Find the mode, median, and mean number of packages sent.

Use the heights in inches of the students in your class to make a table.

15. Make a frequency distribution for the data.

16. Use the table to make a histogram.

17. What is the range of the data?

18. What is the mode of the data?

19. What is the median of the data?

20. What is the mean of the data?

21. Which measure of central tendency best describes the height of the students in your class?

22. Are there any outliers in your data? If so, what effect does the outlier have on the mean of your data?

Mixed Review

Write and solve an equation for each problem.
23. Todd receives a commission of 3% on all sales over $5000. What is Todd's commission on sales of $8500?

24. A local survey shows that 7 out of 12 people wear passenger seat belts. How many people out of 216 passengers would you expect to wear seat belts?

The equation $y = 40x + 100$ models the cost in dollars (y) for renting a backhoe for a number of hours, x.

25. What is the slope of the equation?

26. What is the y-intercept?

27. Graph the equation.

28. What is the value of y when x is 2.5?

LESSON 7.3 SCATTER PLOTS AND CORRELATION

The ABC Video Game Company has recently noticed an increase in the number of defective video games produced. The production manager has asked a quality control technician the reason for this increase. The technician has a hunch that the increase is related to the absentee rate of the workers. She gathers the following data.

Defective Video Games Test 1										
	M	T	W	Th	F	M	T	W	Th	F
Absentee Workers	9	11	5	4	2	7	7	11	10	5
Defective Video Games	9	10	11	6	3	6	8	9	7	4

To check the technician's hunch, plot the data as individual points on a coordinate graph. This graph is called a **scatter plot**.

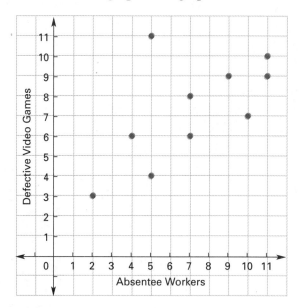

ACTIVITY Information from a Scatter Plot

1 Look at the days that have a low number of absentee workers. Do these days also have a low rate of defective video games?

2 Describe how the points are scattered. Do they appear to cluster around a straight line?

3 Does the line rise or fall as you move from left to right?

4 How does the line represent the relationship between the number of absentee workers and defective video games?

5 Complete Steps 1–4 with the following:

Defective Video Games Test 2									
	M	T	W	Th	F	M	T	W	Th	F
Absentee Workers	7	5	4	5	7	3	7	9	6	2
Defective Video Games	6	4	6	3	4	10	2	3	7	9

6 Complete Steps 1–4 with the following:

Defective Video Games Test 3									
	M	T	W	Th	F	M	T	W	Th	F
Absentee Workers	8	6	8	10	4	3	10	6	5	9
Defective Video Games	6	3	9	8	4	10	4	5	7	5

Correlation

In Test 1 the data show a pattern that resembles a line rising from the left to the right. Since the slope of this line is positive, there is a **positive correlation** between the two sets of data. A positive correlation indicates that both variables in a test are increasing. As the number of absentee workers increases, so does the number of defective video games.

If the slope is negative, a **negative correlation** exists; that is, as one variable increases, the other decreases.

Sometimes there is a positive or negative correlation between data sets. Then you can predict the behavior or trend of one variable if you know the behavior of the other.

If the plot is scattered in such a way that it does not approximate a line, there is no correlation between the sets of data.

What would you conclude about the correlation between the number of defective video games and absentee workers over the six-week period? Are there any other factors than can explain the increase? Just because a positive correlation exists between the two variables does not mean one of the variables *causes* the other to increase.

Ongoing Assessment

Describe the correlation in Test 2 and Test 3. Explain what the correlation means in terms of an increasing or decreasing trend in the data.

Line of Best Fit

A **line of best fit** is a straight line that best represents the data on the scatter plot. This line may pass through some of the points, none of the points, or all of the points.

You can use a measure called the **correlation coefficient**, represented by the variable r, to describe how close the points cluster around the line of best fit. If all the points are on a line and the line has a positive slope, the correlation coefficient r is 1. If all the points are on a line and the line has a negative slope, r is -1. If the data is scattered at random, r is close to zero.

If r is close to 1 or -1, the trend of the data is well represented by a line, and there is a strong relationship between the data and the line. If r is close to zero, there is a weak relationship between the data and the line. In the real world, few relationships show a perfect correlation of 1 or -1.

The scatter plot on the next page shows the relationship of weight (x-axis) to height (y-axis) of the first 12 students to try out for the varsity basketball team. You can use a graphics calculator to plot the data points. In the statistics menu of the graphics calculator, the line of best fit is called the **regression line**. The correlation coefficient from the calculator is about 0.75.

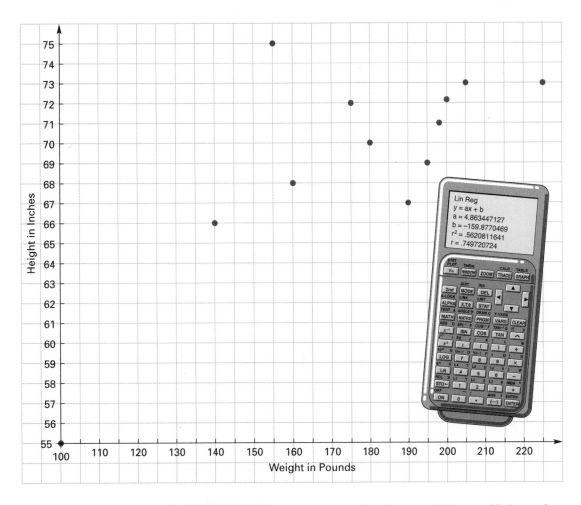

Critical Thinking Explain what the correlation coefficient of 0.75 means with respect to the basketball players.

LESSON ASSESSMENT

Think and Discuss

1 How is data displayed using a scatter plot?

2 Compare the meaning of a positive correlation and a negative correlation between two sets of data.

3 Explain how a "line of best fit" is used with a scatter plot.

Practice and Problem Solving

Jenny works in a state health agency. She is measuring the relationship between the amount of physical exercise per week and age. She records the data as ordered pairs. The first number in each ordered pair is a person's age and the second number is the number of hours of physical activity per week for that person.

(20, 15), (22, 11), (30, 6), (30, 7), (34, 6.1), (26, 13), (26, 8.5), (18, 16), (36, 3), (36, 5.8), (28, 11), (30, 9), (40, 3)

4. Plot the points on a scatter plot.

5. Does the data show any correlation between age and the number of hours of weekly exercise? If so, what kind of correlation?

6. Draw a line of best fit.

7. Use the slope and a point on the line to write an equation to approximate the line of best fit.

8. Enter the data in a graphics calculator and find the regression line.

9. Compare your line and the graphics calculator line and explain any differences.

The relationship between the number of hours studied and the score on an exam is represented by the ordered pairs listed below. The first number of the pair represents the number of hours studied and the second number represents the test score.

(10, 100), (0, 50), (1, 60), (3, 70), (5, 75), (4, 80), (7, 95), (9, 80), (9, 90), (7, 70), (9, 80), (5, 85), (10, 85), (4, 70), (6, 85), (6, 90), (9, 80), (10, 95), (3, 75)

10. Draw a scatter plot for the data.

11. Find the mode of the test scores.

12. What was the test score of the student who studied zero hours?

13. How many students got a 90% on the test? How many hours did they study?

14. Does the scatter plot indicate a negative or positive correlation?

15. Draw a line of best fit.

16. Use the slope and a point on the line to write an equation to approximate the line of best fit.

17. Enter the data on a graphics calculator and find the regression line.

18. Compare your line with the graphics calculator line and explain any differences.

Mixed Review

The fax bill for the Captain Tire Company averages $285 each month. The comptroller asked one of the assistants in the accounting office to compare the fax bills for the last five months. The monthly bills are as follows:

Jan.	Feb.	Mar.	Apr.	May
$240	$210	$325	$290	$305

19. Use integers to show the deviation between the average fax bill and the bill for each month.

20. Draw a number line with $205 at the center point. Use the results of Exercise 19 to place each amount on the number line in the correct order.

21. Boa is pouring a concrete walkway around a circular swimming pool. The radius of the pool is 10 feet. Boa wants the walkway to be 3 feet wide at each point. What is the area of the walkway Boa is planning?

Monique starts walking at a location 2 miles from her office. She walks at a constant speed of 3.5 miles per hour in a direction away from her office.

22. Find a function that models the distance Monique is from her office. Write the function in $f(x)$ notation, where $f(x)$ is distance and x is time.

23. Find $f(x)$ when x is 5.

24. Make a table of ordered pairs for the walk and graph the ordered pairs.

LESSON 7.4 PROBABILITY

Every evening after work, Carlos puts all of his change in a jar. One evening Carlos counts his coins and finds he has the following:

10 quarters   15 dimes   25 nickels

ACTIVITY 1   Part of a Whole

1 How many coins does Carlos have in the jar?

2 What fraction of the coins are

    a. quarters?   **b.** dimes?   **c.** nickels?

3 What percent of the coins are

    a. quarters?   **b.** dimes?   **c.** nickels?

4 Suppose Carlos reaches into the jar and, without looking, removes a coin. What are his chances that the coin will be a quarter?

Removing a coin from a jar is an example of an **event**. In Activity 1, there are 50 ways for an event (removal of a coin) to occur. Each occurrence is referred to as an **outcome**. For example, removing any one of the fifty coins is an outcome.

In Activity 1, Step 4, the outcome of removing a quarter is called the **favorable outcome**. Since there are 10 quarters in the jar of 50 coins, the chance of the favorable outcome (removing a quarter) occurring is 10 out of 50, or 20%.

When Carlos reaches into the jar and removes a coin without looking, he is *randomly* selecting a coin. A random selection means that all coins are *equally likely* to be selected. Whenever all the outcomes are equally likely, the probability is called **theoretical probability**.

Theoretical Probability

The theoretical probability $P(E)$ that an event E will happen is:

$$P(E) = \frac{\text{Number of favorable outcomes}}{\text{Total number of outcomes possible}}$$

Thus, the probability of removing one quarter, $P(\text{Quarter})$ from Carlos' jar of 50 coins is found in the following way:

$$P(\text{Quarter}) = \frac{10}{50} = \frac{1}{5} = 20\%$$

ACTIVITY 2   All or None

Carlos keeps his pennies in a separate jar. So far he has saved 120 pennies. He removes one coin from this jar. What is the probability

1 the coin is a penny?

2 the coin is a dime?

The questions in Activity 2 lead to the following definitions:

Probability of 1 and 0

If an event is sure to happen, the probability is 1.

If an event is impossible, the probability is zero.

Critical Thinking   Complete the following sentence. The probability of any event is greater than or equal to ___?___ but less than or equal to ___?___. Explain why.

A bowl contains 5 red marbles, 3 blue marbles, and 12 green marbles. You remove one of the marbles. What is the probability the marble will be

a. red?   **b.** blue?   **c.** green?   **d.** orange?

e. What is the sum of these four probabilities?

Sondra is a botanist. She is developing new plants for a local nursery. She is crossing plants that have red flowers with plants that have white flowers. A red-flowering plant has two red genes (RR). A white-flowering plant has two white genes (WW). Each parent contributes one gene to the seedlings.

Use a chart to model the possible color combinations for the seedlings. The top row shows the genes for the red-flowering plant. The left column contains the genes for the white-flowering plant. Each seedling has one gene from each parent and produces a pink flower (RW).

		Red-flowering Plant	
		R	R
White-flowering Plant	W	RW	RW
	W	RW	RW

RW = Pink

Sondra now crosses two of the pink-flowering plants. To see what the results might be, Sondra makes a new chart for the second generation.

		Pink Parent Plant	
		R	W
Pink Parent Plant	R	RR	RW
	W	RW	WW

RR = Red
RW = Pink
WW = White

EXAMPLE 1 Probability of an Event

What is the probability that a plant in the second generation is

a. pink?   **b.** red?

SOLUTION

Crossing two pink-flowering plants has the following possible outcomes:

$$(RR) \quad (RW) \quad (RW) \quad (WW)$$

a. There are two ways to get a pink-flowering plant: (RW), (RW). Thus,

$$P(RW) = \frac{2}{4} = \frac{1}{2} = 50\%$$

b. There is one way to get a red-flowering plant: (RR). Thus,

$$P(RR) = \frac{1}{4} = 25\%$$

Since there is one way to get the favorable outcome of a red flower and three ways to fail, the **odds** of getting a red flower are 1 to 3, or $\frac{1}{3}$.

Odds of an Event

$$\text{Odds of an event} = \frac{\text{Number of favorable outcomes}}{\text{Number of unfavorable outcomes}}$$

Critical Thinking How do you find the probability that an event will not happen? What are the odds that an event will not happen?

EXAMPLE 2 Tossing Coins

Suppose you toss two quarters into the air. What are the odds that neither quarter will show heads?

SOLUTION

Heads/Heads

Heads/Tails

Tails/Heads

Tails/Tails

The coins can show (HH), (HT), (TH), (TT). For this example (TT) is the favorable outcome. There is one way to get tails on both quarters and three ways to fail. The odds are 1 to 3 neither quarter will come up heads.

LESSON ASSESSMENT

1 Describe how you can find the probability of buying the winning raffle ticket out of the total number sold.

2 Discuss the circumstances under which the probability of an event is zero or 1.

3 What is meant by the odds of an event occurring?

4 Discuss the difference between probability and odds.

Practice and Problem Solving

A tennis bag contains four white, five orange, and eleven green balls. If the tennis coach takes one ball from the bag, find the probability that it is

5. Orange.   **6.** White.   **7.** Green.

8. The faces of a number cube are numbered 1, 2, 3, 4, 5, and 6. If the cube is tossed, what are the odds that a 5 will turn up?

9. Past records show that a machine's probability of producing a defective bearing is 12%. Find the probability that a bearing from that machine will not be defective.

10. A store running a sales promotion has a box containing 60 envelopes. Twenty of the envelopes hold $5 bills. The other envelopes are empty. If one envelope is drawn, what is the probability that it will contain $5?

The chart gives a current inventory of T-shirts. If you choose one shirt from the inventory, what is the probability that it will be

	Large	Small
Blue	11	6
Red	14	9
Green	5	10

11. a red T-shirt?

12. a small T-shirt ?

13. a large, green T-shirt?

14. Write a problem in which the probability of an event happening is $\frac{3}{4}$.

15. Write a problem in which the odds of an event occurring are $\frac{2}{3}$.

The personnel office randomly schedules two employees, Leo and Vince, to work during two shifts, A and B.

16. Write all the possible outcomes for shift assignments as ordered pairs. For example, let (LA, VB) represent Leo on the first shift and Vince on the second shift.

17. What is the probability that Leo and Vince will be assigned to the same shift?

18. What is the probability Vince will be assigned to shift A?

Suppose you toss a penny, a nickel, and a dime. A head on the penny, a tail on the nickel, and a head on the dime can be recorded as the ordered triple (h, t, h). There are eight possible ways to record the heads and tails of a toss.

19. Use ordered triples to indicate all the possible outcomes of tossing the three coins.

20. What is the probability that all the coins will land heads up?

21. What are the odds that only one of the coins will land heads up?

22. A rectangular garden is 12 feet by 16 feet. A path is built diagonally from one corner to the opposite corner. How long is the path?

Write and solve an equation for each situation.

23. Each base angle of an isosceles triangle is 30 degrees less than the vertex angle. Find the measure of the vertex angle.

24. A board 600 centimeters long is cut into two pieces. One piece is three times the length of the other. How long is the shorter section?

25. The sum of three consecutive integers is -57. What are the integers?

The function $f(x) = x^2$ is an example of a nonlinear function.

26. Graph f for the values of x such that $0 \leq x \leq 3$.

27. If the graph of f is translated two units to the right, what are the new coordinates of the vertex?

28. How is the graph of $g(x) = (x + 1)^2$ related to the graph of f?

LESSON 7.5 EXPERIMENTAL PROBABILITY

In Lesson 7.4 you used theoretical probability to predict the outcome of events where all the outcomes were equally likely to occur. For example, when you toss a coin there is a theoretical probability of 50% that it will come up heads.

However, not all events are random. Assume there are 10,000 cars in your city and last night an accident occurred. You would not expect that the probability of being in the accident to be $\frac{1}{10,000}$ for each car in the city. In this situation, the probability depends on many factors, such as the weather and how other people drive.

When events are not random, you use **experimental probability**. In experimental probability, sample data or observations are used to estimate the probability of a specific event occurring. For example, statisticians for auto insurance companies collect data to determine the probability that people of various ages will have an accident. These probabilities are the key factors in setting insurance rates.

You can calculate experimental probability for both random and nonrandom events.

ACTIVITY 1   An Experiment with a Penny

1 Make a table like this for 20 trials.

Trial	Heads	Tails
1	?	?
2	?	?
3	?	?

2 Toss a penny 20 times. Record how the coin comes up each time. For each trial, record *H* under the Heads column and *T* under the Tails column.

3 How many times did the coin come up heads? What percent of the tosses were heads?

4 If you toss a penny 10 more times, how many times do you think it will come up heads?

Heads   Tails

5 Try 10 more tosses and record the result. Is the result the same as your guess?

6 Based on your results of this experiment of randomly tossing a penny, what is the probability that you will get 5 heads if you toss a penny 10 times?

Activity 1 shows how data is used to find experimental probability. The experimental probability of an event is found by collecting data and evaluating the data to predict the outcome of a specific event.

Usually, you observe a large number of trials and count how many times the event occurs. Once the count is completed, you can use the following formula to calculate the experimental probability.

> **Experimental Probability**
> Based on sample data or observations, the experimental probability $P(E)$ that an event E will happen is
>
> $$P(E) = \frac{\text{Number of times the event happens}}{\text{Total number of tries}}$$

Ongoing Assessment

Set up an experiment with a number cube that has one number (1 through 6) on each face. What is the experimental probability of rolling a 4 in one try? Compare your results with the theoretical probability that the same event will occur.

As you found in Activity 1, you can use experimental probability to predict the outcome of a random event. However, experimental probability is most useful when the possible outcomes are *not* equally likely.

EXAMPLE 1 Make a Prediction

A consumer group surveyed its members and found that many of them had problems with a certain brand of tire. Out of a total of 14,537 tires purchased, 422 developed problems within the first 1000 miles. Unfortunately, you have just had one of these tires installed on your car. What is the probability that your tire will have a problem in the first 1000 miles of use?

Solution

The only data you have is experimental. Thus, use the experimental probability formula.

$$P(\text{tire failure}) = \frac{422 \text{ tire failures}}{14{,}537 \text{ tires purchased}} \approx 2.9\%$$

Using experimental probability, you can predict that there is about a 2.9% chance (or about 1 chance in 34) that you will have a tire problem in the first 1000 miles.

Random Numbers

In **inferential statistics**, you use random sampling of a group to predict trends or behaviors of the group. Random sampling assumes that every possible sample of the group has the same probability of being selected.

Under these circumstances, you can use **random numbers** to choose sample selections. Calculators or spreadsheet programs can generate random numbers. Many statistics books contain lists of randomly generated numbers.

EXAMPLE 2 A Random Number Problem

What is the probability that a pair of random numbers ranging in value from 1 to 9 contains a 3?

Solution

Generate 20 random numbers from 1 through 9. Use the numbers to form ten pairs. Count the number of pairs that contain a 3.

The table lists ten pairs of random numbers using the numbers from 1 through 9. In three of the pairs, one of the numbers is a 3. Thus, for this sample the experimental probability is $\frac{3}{10}$, or 30%.

1,6	2,2
2,1	4,**3**
3,6	7,1
9,7	9,9
8,5	2,**3**

Critical Thinking   In Example 2 find the experimental probability that at least one pair will add up to 10 or less.

Simulations

When it is impossible to set up an experiment to test an entire event you can sometimes use a **simulation** instead of the real experiment.

Suppose a potential wholesale supplier of tomato seeds for your garden supply store claims that half its seeds germinate within ten days of being planted. To verify the claim, you plant 20 seeds according to instructions. At the end of ten days, you dig up the seeds and test them for germination. The first four seeds you test have not germinated. If the supplier's claim is correct, what is the probability of having dug up four nongerminated seeds in a row?

Activity 2 presents a way of using random numbers to answer this question. The coin flipping simulates the germination rate of tomato plants.

ACTIVITY 2   A Coin Toss Simulation

1 Look at your record in Activity 1. Write your data so that the results of the 20 coin tosses are in the order they happened. For Example, H, H, T, H, ...

2 Examine the sequence to see if there are any places where four heads in a row occur.

3 How many times does the sequence H, H, H, H occur? Write a ratio using this number as the numerator and 20 as the denominator. What experimental probability did you get?

4 Combine your data with the rest of your class. Calculate the experimental probability from the larger data set. How does the probability from the larger set of data compare with that of your experiment with twenty data points?

There are many ways to generate random numbers to simulate experiments. You can use coins, number cubes, spinners, calculators, computers, or tables. The important thing to remember is that the more trials you take, the closer the simulation approximates the results of random events in the real world.

CULTURAL CONNECTION

Every culture plays games where winning is decided by the luck of the toss or the luck of the draw. Early American Indian children played a game that consisted of tossing five flat sticks with pictures on one side. After the toss, points were awarded based on the number of sticks showing the pictures.

In the seventeenth century, two famous French mathematicians, Pierre Fermat and Blaise Pascal, were given a problem similar to the one that follows.

The Situation:

Two players with equal skills are involved in a game of chance. The game is stopped before the game can be completed. Player A needs two points to win and Player B needs three points to win.

The Problem:

How should the amount of money already wagered be split between Player A and Player B?

WORKPLACE COMMUNICATION

You need to do some preplanning before you set up the equipment for the $P(BE)$ verification test: Assume the CELSAT data communications system achieves the specified $P(BE)$. For this probability of error, how many bits (tries) will be required to record the required 10^4 errors (events) in the verification test? At the data rate of the commercial modem, how many hours do you need to schedule to complete the test?

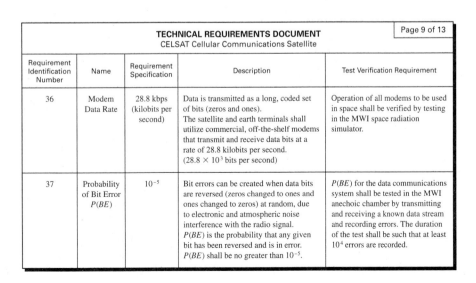

			TECHNICAL REQUIREMENTS DOCUMENT CELSAT Cellular Communications Satellite		Page 9 of 13
Requirement Identification Number	Name	Requirement Specification	Description	Test Verification Requirement	
36	Modem Data Rate	28.8 kbps (kilobits per second)	Data is transmitted as a long, coded set of bits (zeros and ones). The satellite and earth terminals shall utilize commercial, off-the-shelf modems that transmit and receive data bits at a rate of 28.8 kilobits per second. (28.8×10^3 bits per second)	Operation of all modems to be used in space shall be verified by testing in the MWI space radiation simulator.	
37	Probability of Bit Error $P(BE)$	10^{-5}	Bit errors can be created when data bits are reversed (zeros changed to ones and ones changed to zeros) at random, due to electronic and atmospheric noise interference with the radio signal. $P(BE)$ is the probability that any given bit has been reversed and is in error. $P(BE)$ shall be no greater than 10^{-5}.	$P(BE)$ for the data communications system shall be tested in the MWI anechoic chamber by transmitting and receiving a known data stream and recording errors. The duration of the test shall be such that at least 10^4 errors are recorded.	

LESSON ASSESSMENT

1 Compare experimental probability with theoretical probability.

2 Explain how random numbers can be used to simulate a probability problem.

Practice and Problem Solving

Toss a coin 100 times. Record each toss as heads (H) or tails (T).

3. What is the difference between the number of heads and the number of tails?

4. Record the total number of heads and tails tossed by your class.

5. Using theoretical probability, how many heads and how many tails would you expect for the data from the entire class?

6. Compare your class's experimental results and the theoretical probability.

Toss a nickel and a penny 100 times and record the heads (H) and tails (T) as ordered pairs. Record the nickel as the first number in the pair.

7. What does (T, H) represent?

8. What is the experimental probability that both coins will come up heads on a single toss?

9. List the four possible outcomes as ordered pairs.

10. What is the theoretical probability that both coins come up heads?

11. Compare the theoretical probability and the experimental probability.

12. Based on your experimental results, how many times would you expect two heads to come up if you toss the two coins 100 more times?

Roll a number cube 100 times. Record each number as it comes up.

13. Use your data to estimate the probability of rolling a 6 on a single roll.

14. Based on your results, how many 4s would you expect to get in another 100 tosses?

15. Combine the class results from tossing a number cube. Use this data to estimate the probability of rolling a 6 on a single roll.

16. What is the theoretical probability that a 6 will come up on one roll of the number cube? Compare the theoretical probability to the experimental probability.

Describe a simulation using random numbers to estimate the answer to each of the following situations.

17. Music practice is held once a week, Monday through Friday. What is the probability it will be held on Wednesday?

18. You take a true-false test with 5 questions. What is the probability of guessing 4 questions correctly?

19. You have a 50-50 chance to make a free throw. What is the probability you will make 2 in a row if you take 5 shots?

Mixed Review

20. A storage tank holds 9000 cubic meters of water. Use scientific notation to write the volume of the tank in cubic centimeters.

Write and solve an equation for each situation.

21. Paul is responsible for mowing 40 miles of highway medians for the state. If he has completed $\frac{2}{5}$ of his mowing, how many miles does he have left to mow?

22. In a random sample of 400 voters, 90 people said they planned to vote in the next city election. If there are 10,000 registered voters in one district, how many of those people can be expected to vote?

23. Becky is a quality control supervisor. On Monday she found several defective parts. On Tuesday she found 25 more defective parts than she did on Monday. If Becky found 67 defective parts altogether, how many did she find on Monday?

24. The distance around a rectangular room is 72 feet. If the length of the room is 16 feet, what is the width?

25. An airplane is flying at an altitude of 25,000 feet. To land, the plane must descend at a rate of 750 feet per second. Write an equation for the altitude (in feet) as a function of time (in seconds) to model the descent of the airplane from its current altitude.

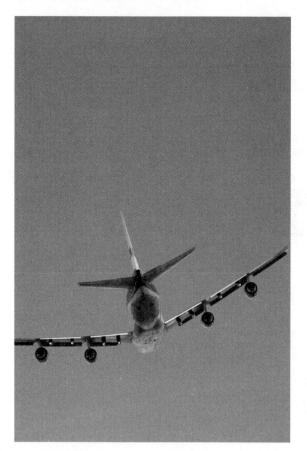

LESSON 7.6 THE ADDITION PRINCIPLE OF COUNTING

Suppose you roll a red number cube and a green number cube. What is the probability that a red 3 and a green 4 will come up?

Probability of A and B

Since any of the six faces of a number is equally likely to come up, use theoretical probability. The probability that you toss a red 3 and a green 4 is the ratio of the number of favorable outcomes to the number of the possible outcomes.

Use a set of ordered pairs to represent all the possible outcomes of rolling different colored number cubes. You can graph these outcomes on a coordinate system. Use the x-axis to represent the possible outcomes for the red cube. Use the y-axis to represent the possible outcomes for the green cube.

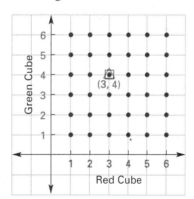

Each ordered pair on the graph represents one possible outcome of the event. The entire set of possible outcomes is called a **sample space** for the event.

The sample space contains 36 possibilities for tossing a red cube and a green cube. But there is only one way to toss a red 3 and a green 4. Thus,

$$P(\text{R3 and G4}) = \frac{1}{36}$$

Probability of A or B

What is the probability that a red 3 *or* a green 4 will occur in one toss of the cubes?

In probability, the words *and* and *or* have special meanings. The word *and* means both events occur. The word *or* means one of the events or both of the events occur.

The ordered pairs representing a red 3 are graphed using a square. The ordered pairs representing a green 4 are graphed using a triangle. There are 11 ordered pairs representing a toss that contains a red 3 or a green 4.

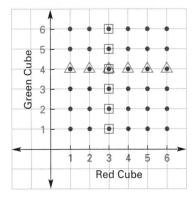

Thus, the probability of tossing a red 3 or a green 4 is

$$P(\text{R3 or G4}) = \frac{11}{36}$$

or about 31%.

Be careful when you count the events on a graph. The combination of a red 3 and a green 4 is a single event and should only be counted once.

EXAMPLE 1 *P*(Sum 7)

If you roll the two number cubes once, what is the probability that the sum of the numbers will be 7?

SOLUTION

Make a sample space for the cube toss. Circle all the points that represent a sum of 7. Since there are 6 such points.

$$P(\text{sum 7}) = \frac{6}{36} = \frac{1}{6}$$

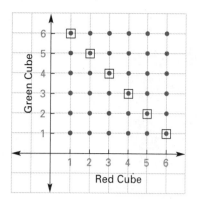

Critical Thinking What is the probability of tossing a sum of 5 or less on one toss of the two number cubes.

Counting Choices

EXAMPLE 2 Art and Music

The chart shows enrollment for art and music classes at a local high school.

	Music	Art	Total
Boys	10	18	28
Girls	15	12	27
Total	25	30	55

a. How many students are boys?

b. How many students are in music?

c. How many students are boys *and* are in music?

d. How many students are boys *or* are in music?

SOLUTION

a. The first row shows 28 students are boys.

b. The first column shows 25 students are in music.

c. The intersection of the first row and first column shows 10 boys are in music.

d. 28 of the students are boys. There are 25 students in music. If you add 28 and 25, then you must subtract 10, because 10 of the boys have been added twice. Thus, 43 students are boys or are in music.

Use the chart from Example 2. Find the number of students who are girls or are in art.

The examples lead to an important principle of counting.

Addition Principle of Counting

Suppose there are *m* ways to make a first choice, *n* ways to make a second choice, and *s* ways to count the choices twice. Then the number of ways to make the first choice *or* the second choice is

$$m + n - s$$

The Addition Principle of Counting applies to the probability problems in this lesson. For example, there are six combinations in which the red cube can come up 3. There are six combinations in which the green cube can come up 4. But as stated before, the one combination of a red 3 *and* green 4 is counted twice. Thus, there are $6 + 6 - 1$ (or 11) favorable outcomes for tossing a red 3 or a green 4 out of 36 possible choices.

$$P(\text{R3 or G4}) = \frac{11}{36}$$

Mutually Exclusive Events

If two events cannot occur at the same time, the events are **mutually exclusive**. If you remove a coin from a jar of coins, you cannot pick one coin that is *both* a quarter and a dime. You can use the fact that the results are mutually exclusive to determine the probability that either one event or the other event will occur.

Recall how the coins in Carlos' jar are distributed.

10 quarters	15 dimes	25 nickels

If Carlos removes one coin from the jar, what is the probability that the coin is either a quarter or a dime? There are too many events to build a sample space and graph the favorable outcomes. Instead, look at the probability that each event can happen and find the sum of those probabilities.

The probability of removing a quarter is $\frac{10}{50}$.

The probability of removing a dime is $\frac{15}{50}$.

Thus, the probability of removing a quarter or a dime is

$$\frac{10}{50} + \frac{15}{50} = \frac{25}{50}, \text{ or } 50\%$$

The probability of removing a quarter or a dime is 50%. If Carlos returns the coin to the jar and repeats the experiment 100 times, he will get either a quarter or a dime on about 50 of the tries.

> **Probability of Mutually Exclusive Events**
> If A and B are mutually exclusive events,
> $$P(A \text{ or } B) = P(A) + P(B)$$

LESSON ASSESSMENT

Think and Discuss

1 What is a sample space?

2 Explain the difference between the use of *and* and *or* in finding probabilities.

3 When is the Addition Principle of Counting used in finding probability?

4 What are mutually exclusive events?

Practice and Problem Solving

Refer to the sample space for rolling two number cubes. Find the probability of each event.

Red Cube / Green Cube scatter grid

5. A red 4 and a green 6.   **6.** A red 4 or a green 6.

7. A red prime number and a green even number.

8. A red composite number or a green multiple of 2.

9. Either cube shows a 1.   **10.** Neither cube shows a 1.

11. Both cubes show a 1.   **12.** The sum is 5.

13. The sum is less than 7.   **14.** The sum is greater than 12.

15. There are six entrances to a building. Find the probability that two firefighters, entering at random and at different times, will choose the same door.

Four tickets in a box are numbered from 1 to 4. If two tickets are drawn, find the probability that:

16. the sum of their numbers is odd.

17. the product of their numbers is even.

The chart gives the current inventory of T-shirts. If you choose two shirts from the inventory, what is the probability that the T-shirts will be

Color	Large	Small
Blue	11	6
Red	14	9
Green	5	10

18. red and small?   **19.** blue or large?   **20.** large and not red?

21. The probability that you will buy an American brand of car is $\frac{3}{5}$. The probability that you will buy a German brand of car is $\frac{1}{4}$. What is the probability that you will buy an American or German brand of car?

22. The probability that it will rain is 0.6. The probability that it will be partly cloudy is 0.15. What is the probability that it will rain or be partly cloudy?

23. Solve the formula $i = prt$ for r.

24. A wire 24 inches long is bent to form a rectangle. If $x + 2$ represents the length of the rectangle, write an expression for its width in terms of x.

25. Mario is the supervisor for production in a cellular phone plant. In May Mario's shift made 200 more phones than in April. In June the number of phones made was double that of May. If Mario's shift production totaled 3800 items, what were each months production figures?

26. A garden is in the shape of a triangle with a perimeter of 37 feet. The longest side of the triangle is 5 feet greater than the shortest side. The remaining side is 2 feet greater than the shortest side. Find the dimensions of the garden.

LESSON 7.7 THE FUNDAMENTAL COUNTING PRINCIPLE

Suppose you have the following clothes in your closet.

Shirts: plaid, red, blue, tan
Pants: brown, black
Shoes: plastic sandals, canvas shoes, leather shoes

In how many different ways can you choose the three articles of clothing?

You can answer a question by drawing a **tree diagram**.

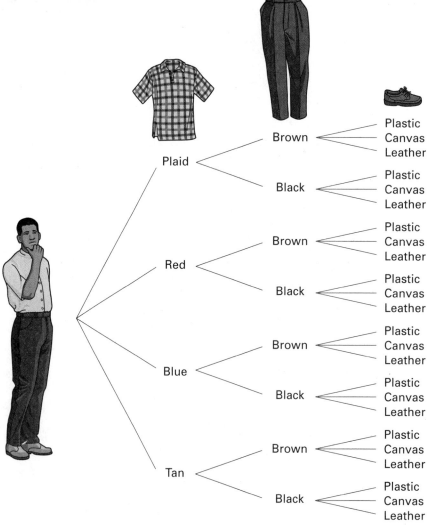

Start at the left side of the tree and follow one path all the way to the right. For example, start by choosing the plaid shirt. You can choose either black or brown pants. If you choose the brown pants, you can choose either plastic, canvas, or leather shoes. If you choose the black pants, you can choose either the plastic, canvas, or leather shoes. Thus, there are 6 ways to choose a combination of clothing if you start with a plaid shirt.

Ongoing Assessment

Begin with a different shirt and follow the paths to the right. How many different paths can you take?

The easiest way to count all possible paths is to count the final branches at the right side of the tree diagram. The final count for all the possible clothing combinations is 24.

ACTIVITY 1   Making Choices

To order a radio, you can make one choice from each of these three categories:

Colors: green, pink, yellow
Type: AM, FM, AM/FM
Accessories: battery, tape
    player

1 Create a tree diagram to show the possible combinations of color, type and accessory.

2 How many combinations are there?

Creating a tree diagram is a good way to picture choices. However, it takes time, and if there are many choices, there may not be enough room on your paper. Another way to organize and count the possible combinations in the clothing problem is found in Activity 2.

ACTIVITY 2   Multiplying to Find Choices

1 Draw three boxes, one for each kind of clothing.

Shirts	Pants	Shoes

2 In the first box, write the number of choices you can make in selecting a shirt. There are 4 (plaid, red, blue, and tan). Complete the boxes for the other possible choices.

3 Find the product of the numbers in the three boxes. How does this product compare to the number of combinations you found using the tree diagram?

Ongoing Assessment

Use the method in Activity 2 to find the number of choices in Activity 1. Compare the results of the two methods.

Activity 2 is an application of the **Fundamental Counting Principle**.

> ### Fundamental Counting Principle
> To find the number of ways to make successive choices in a series of different categories, multiply together the number of choices in each category.

For every choice of shirt in the clothing problem, there were two choices of pants. For every choice of pants, there were three choices of shoes. A choice in one category of clothing did not affect the number of choices in any other category.

Similarly, in the radio problem, a choice of color, type, or accessory did not affect the number of choices in either of the other two categories. However, in some problems, a choice in one category eliminates choices from other categories.

EXAMPLE 1 Filling Officer Positions in a Club

The computer club elects three officers—president, secretary, and treasurer. Five members of the club (Akers, Chen, Miller, Vincent, and Wohl) are eligible for election as officers. How many different ways can these 5 people fill the 3 positions?

SOLUTION

For president (first position) there are 5 choices (Akers, Chen, Miller, Vincent, and Wohl). The first position can be filled in 5 different ways.

After one person is chosen to fill the first position, there are only 4 choices left to fill the second position of secretary. The second position can be filled in 4 different ways.

For treasurer (third position) there are only 3 choices left. The third position can be filled in 3 different ways. Use the Fundamental Counting Principle to find the number of ways to choose.

$$\begin{array}{ccccccc} \text{President} & & \text{Secretary} & & \text{Treasurer} \\ 5 & \bullet & 4 & \bullet & 3 & = 60 \end{array}$$

There are 60 different ways to fill the 3 positions given 5 choices.

EXAMPLE 2 Sitting at a Table

If 5 people can sit at the head table for a banquet, and there are 10 people to choose from, how many different seating arrangements can there be?

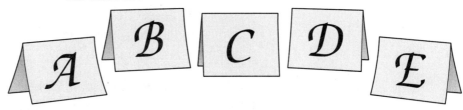

SOLUTION

Draw a picture to show the 5 positions. There are 10 ways to fill position A. After A is filled, there are 9 ways to fill position B. Use the Fundamental Counting Principle.

$$10 \bullet 9 \bullet 8 \bullet 7 \bullet 6 = 30{,}240$$

Thus, there are 30,240 possible ways for 5 of the 10 people to sit at the head table.

LESSON ASSESSMENT

Think and Discuss

1 Explain how to make a tree diagram to find the total number of ways to make two choices if each choice has three different possibilities.

2 When is it easier to use the Fundamental Counting Principle rather than to create a tree diagram? Give an example.

In each of the following situations, use the Fundamental Counting Principle. Then, check your work by drawing a tree diagram.

3. If Nancy has 3 skirts, 2 blouses, and 4 scarves, how many different outfits can she wear?

4. A penny, a nickel, and a dime are tossed. How many different outcomes are possible?

5. You are placing decals on model race cars. Your choices consist of blue, white, green, or black decals in sizes small, medium, and large. How many choices do you have?

6. How many ways can you choose a team captain and a co-captain from a group of 5 players?

7. The day care center at a semiconductor plant serves breakfast and lunch Monday through Friday. How many meals are served each week?

8. Brian plays clarinet and saxophone, Hector plays the oboe, and Gina plays piano, flute, and trumpet. If Brian, Hector, and Gina form a trio, how many combinations of instruments can they play?

Use the Fundamental Counting Principle to answer the following questions.

9. Tickets for 10 performances of a play are sold for 6 different prices. How many combinations of performances and prices must be printed?

10. While Jeff is buying new tires for his car, he finds that they come in 16 different sizes, 4 different grades, 3 different types of construction, and either white sidewall or black sidewall. How many tire choices does Jeff have?

11. The Down Home Cafe serves 3 kinds of soup, 3 kinds of salad, and 5 choices of salad dressing. How many different orders of soup and salad (with dressing) are possible?

12. How many outfits are possible with 4 shirts, 2 pairs of slacks, 4 sweaters, and 3 ties?

13. There are 6 members of a club. How many ways can they elect a president and a secretary?

14. There are 5 runners in the 100-meter dash. Find the number of different ways the runners can finish the race in first through fifth place.

15. Seven students are in the finals of a skateboard competition. Only first, second, and third places are awarded. How many ways can the students finish in the competition?

16. Each school lunch package contains a ham, a chicken, or a turkey sandwich and a lemonade or orange drink. If the packages are randomly distributed, what is the probability that you will get a chicken and lemonade combination?

Mixed Review

Write and solve an equation for each situation.

17. Amber has saved $\frac{3}{5}$ of the cost for her first year at community college. If Amber has saved $690, what is the first year cost of college?

18. Ann and Kerry worked a total of 42 hours at their part-time jobs. Ann noticed that she worked 6 more hours than Kerry. How many hours did Kerry work?

19. Gilberto works part-time and Kyle works full-time. Kyle earns four times as much as Gilberto. If the difference in their pay is $150, how much does Gilberto earn?

20. The length of one side of a triangle is $\frac{1}{2}$ that of the second side. The length of the third side is $\frac{4}{5}$ the length of the second side. If the perimeter of the triangle is 138 inches, what is the length of each side?

Graph each function.

21. $f(x) = |x|$ **22.** $h(x) = |x| + 1$ **23.** $g(x) = |x| - 1$

24. Give the coordinates of the vertex for each absolute value function.

LESSON 7.8 INDEPENDENT AND DEPENDENT EVENTS

Margo has a bag that contains two red marbles, a blue marble, and a green marble. She draws one marble at random from the bag. She replaces the marble and draws another marble from the bag.

The probability that Margo removes a red marble on her first draw is $\frac{1}{2}$. Since she replaced the marble, the probability of removing a red marble on the second draw is still $\frac{1}{2}$. This is an example of events that are **independent**. Two events are independent if the outcome of the first event does not affect the second event.

ACTIVITY Multiplying Probabilities

1 If a quarter is tossed once, what is the probability it will come up heads?

2 If a dime is tossed once, what is the probability it will come up heads?

3 Make a sample space for tossing a quarter and a dime.

4 Use the sample space to find the probability both tosses are heads.

5 What is the product of the probabilities from Steps 1 and 2? How does this result compare with your result from Step 4?

Tossing a quarter and tossing a dime are independent events. In the Activity, you showed that the probability of two independent events can be found by multiplying the probabilities of each event.

> **Probability of Independent Events**
> If A and B are independent events, then
> $$P(A \text{ and } B) = P(A) \cdot P(B)$$

For Margo, this means that the probability of removing a red marble on the first draw and removing a red marble on the second draw is
$$\frac{1}{2} \cdot \frac{1}{2} = \frac{1}{4}$$

EXAMPLE 1 Number Cubes

A red number cube and a green number cube are tossed. What is the probability that the number on the red cube is even and the number on the green cube is less than 5?

SOLUTION

Tossing an even number on the red cube and a number less than 5 on the green cube are independent events.

The probability that the red cube comes up with an even number is 3 out of 6, or $\frac{1}{2}$. The probability that the green cube comes up with a number that is less than 5 is $\frac{4}{6}$, or $\frac{2}{3}$.

Since the tosses of the red cube and the green cube are independent events,

$$P(\text{even red and green} <5) = \frac{1}{2} \cdot \frac{2}{3} = \frac{1}{3}$$

Check this answer by using a sample space of ordered pairs.

Ongoing Assessment

Eight cards are numbered 1, 2, . . ., 8. The cards are placed in a bag. One card is drawn and replaced. Then another card is drawn. What is the probability that the first card is greater than 5 and the second card is divisible by 2?

Suppose Margo does not replace the first marble before she draws a second marble. If Margo removed a red marble on the first draw there are 3 marbles left and only one of them is red. On the second draw, the probability that Margo will draw a red marble is $\frac{1}{3}$. This is an example of events that are **dependent**. Two events A and B are dependent when the first event A affects the probability of the second event B. If B follows A, then the probability of B occuring is written $P(B|A)$ and is read, "the probability of B *given that* A has occurred."

Probability of Dependent Events

If A and B are dependent events, and B follows A, then

$$P(A \text{ and } B) = P(A) \cdot P(B|A)$$

Thus, for Margo the probability of removing a red marble on the first draw and a red marble on the second draw is

$$\frac{1}{2} \cdot \frac{1}{3} = \frac{1}{6}$$

EXAMPLE 2 Dependent Events

You have a bag containing 3 red marbles, 2 blue marbles, and 5 green marbles. You draw one marble *but do not replace it*. Then you draw another marble. What is the probability of drawing a red marble on the first draw and a green marble on the second draw?

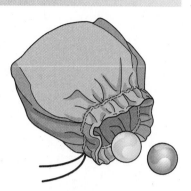

SOLUTION

For the first draw, there are 10 marbles in the bag. Each of the 3 red marbles is equally likely to be drawn. Thus,

$$P(\text{red on first draw}) = P(A) = \frac{3}{10}$$

After the first marble is drawn, there are 9 marbles left. If a red marble was taken on the first draw, there are 2 red marbles, 2 blue marbles, and 5 green marbles left in the bag.

$$P(\text{green on second draw}) = P(B|A) = \frac{5}{9}$$

Since removing a red marble and then removing a green marble without replacing the first red marble are dependent events,

$$P(A \text{ and } B) = P(A) \cdot P(B|A)$$

$$P(\text{red on first draw and green on second draw}) = \frac{3}{10} \cdot \frac{5}{9} = \frac{1}{6}$$

Ongoing Assessment

Recall the coins in Carlos' jar: 10 quarters, 15 dimes and 25 nickels. Suppose Carlos removes 3 coins from his jar without replacing any of them. What is the probability that he will remove a quarter on the first draw, a dime on the second draw, and a nickel on the final draw?

LESSON ASSESSMENT

1 What is the difference between a dependent event and an independent event? Give an example.

2 How can you find the probability of tossing a two on a red number cube and a different number on a green cube

a. using a sample space?

b. using the independent events formula?

Practice and Problem Solving

Use these two spinners. What is the probability you will spin

 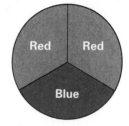

Spinner 1   Spinner 2

3. a 4 and a blue?

4. a 2 and a red?

5. a 1 or a red?

6. a prime number and a blue?

7. a number greater than 3 or a red?

Three coins are tossed. What is the probability that

8. 3 heads come up?

9. exactly two heads come up?

10. exactly 1 head comes up?

11. no heads come up?

12. Why must the answers to Exercises 8 and 11 be the same?

A quality control schedule for bicycle gears requires an accuracy check of a sample consisting of three large gears, four intermediate gears, and five small gears. In the first check, one gear is pulled out of the sample and checked. Then a second gear is pulled out and checked. What is the probability that

13. both gears are small?

14. the first gear is small and the second is large?

In the second check, one gear is pulled and then replaced. Then the second gear is pulled. What is the probability that

15. both gears are small?

16. the first gear is small and the second is large?

Circle B is drawn inside circle A. The area of circle B is $\frac{1}{4}$ the area of circle A. Suppose someone has randomly placed two dots within circle A. What is the probability that

17. both dots are in circle B?

18. one dot is inside circle B and one dot is outside circle B?

19. both dots are outside circle B?

20. both dots are inside circle A?

21. both dots are outside circle A?

The letters $M\ A\ T\ H\ E\ M\ A\ T\ I\ C\ A\ L$ are written on notecards and placed in a hat. One card is drawn and not replaced. Then a second card is drawn. What is the probability that

22. the first card is a T and the second is an M?

23. the first card is an A and the second is an A?

24. the first card is a vowel and the second is a consonant?

25. the cards can be used to spell AT? (Hint: This includes an *or* situation.)

26. A three-person production team is chosen randomly from a group of three men and three women. What is the probability that the team will consist of three women and one man?

27. The top four sales representatives, Mark, Samantha, Rosa, and Ron, are in a random drawing. Only first and second prize will be awarded. What is the probability that Mark will win first prize and Rosa will win second prize?

Mixed Review

Alicia sells lamps at craft fairs. She uses the function $f(x) = 25x - 250$ to project her profit for selling x lamps.

28. Graph the function and identify the slope and y-intercept.

29. At what point does Alicia begin to make a profit?

30. If Alicia makes a profit of $125, how many lamps did she sell?

Write and solve an equation for each situation.
31. A certain new car depreciates at the rate of $2400 per year. A new car that cost $26,000 is now worth $16,400. How old is the car?

32. A plumber needs to buy three sections of pipe for a new home construction project. The length of the longest pipe is twice the length of the shortest pipe. The other pipe is one-and-three-fourths the length of the shortest pipe. In all, the plumber needs 57 feet of pipe. How long is each section of pipe?

MATH LAB

Equipment   Calculator
Meter Stick

Problem Statement

A reaction time is how long it takes to do something (such as step on the brakes) after you receive a certain signal (such as seeing a child start across the street in front of you). Reaction times vary from one person to the next. Reaction times, from one measurement to the next, also vary for the same person.

You will measure the reaction time for stopping a meter stick that falls between your thumb and forefinger. Based on the data for your group and your class, you will calculate measures of central tendency, make a frequency distribution table, and draw a histogram.

Procedure

a One person should hold the meter stick at the 100 cm end. Another member of your group (whose reaction time is being measured) should place his or her hand at the bottom of the stick with thumb and forefinger each two centimeters to either side of the 0 cm mark. When the person holding the meter stick drops it, the other person, looking only at the bottom of the stick, catches the stick as quickly as possible by pressing thumb and forefinger together.

b The length of the meter stick from the 0 cm end to the "catch point" is related to the elapsed time from the moment of "drop" to the moment of "catch." Thus, you can use this length as a measure of the reaction time. For each drop, read the millimeter mark that is just visible above the thumb. Use this reading as an indication of the reaction time. A lower reading (75 mm) represents a faster reaction time than a higher reading (92 mm).

c Have each member of the group catch three drops. Record the meter stick readings for each drop.

d When all groups have completed their measurements of reaction times, write the meter stick readings for the entire class on the chalkboard. These readings become the data set for the class.

e Divide the reaction times for the class data into five or six convenient intervals (for example, 50–59 mm, 60–69 mm, 70–79 mm, and so on). Make a frequency distribution table.

f Draw a histogram of the class data.

g Determine the mode and the median of the reaction times for the entire class.

h Find the mean of the class data.

i Compare your group data to the class histogram.

Activity 2: Games That Use Dice

Equipment   Calculator
Two dice

Problem Statement

You will examine the possible outcomes that occur from throwing a single die and a pair of dice. Many board games, such as Monopoly™, use dice to advance players around the playing board. On any throw, each face on a die is equally likely to come up. Since only one face can come up at a time, the results are mutually exclusive.

When you throw two dice, the sum of the dots on the faces can be anything from 2 through 12. These sums are not equally likely. Even though there is only one way to throw a 2 (1+1), there are six ways to throw a 7 (1+6, 2+5, 3+4, 4+3, 5+2, and 6+1). Thus, the events throwing a 2 and throwing a 7 are not equally likely.

Procedure

a Have each member of the group throw a single die 12 times and a pair of dice 12 times. Record the results. For a single-die throw, record the dots on the face that come up. For a two-die throw, record the sum of the dots on the two faces that come up. For example, if you throw a 5 and a 3, record an 8.

b Make a table similar to that shown below for the data you collect. The single-die throw should list the "face showing" and the "corresponding frequency" for the 12 throws. The two-dice throw should list the "sum of the two faces" and the "corresponding frequency," also for 12 throws.

Single-die Throw		Two-dice Throw	
Face Showing	Corresponding Frequency	Sum of Faces	Corresponding Frequency
1	2	2	0
2	3	3	0
3	1	4	3
4	2	5	2
5	2	6	1
6	2	7	2
		8	1
		9	1
		10	1
		11	1
		12	0

c Draw histograms for both the single-die and two-dice data.

d Make a table similar to the one above for all your group's data. Draw histograms of the single-die data for the group and the two-dice data for the group.

e Make a data table for the entire class. Draw separate histograms of the single-die data and the two-dice data for the class.

f Are there similarities in the three histograms for the single-die data? Are there similarities in the three histograms for the two-dice data? How are the histograms of single-die data different from the histograms of the two-dice data?

Equipment   Calculator

Problem Statement

You will examine the frequency distributions for three physical traits. These traits are hair color, eye color, and dominant hand (left or right). Based on these experimental probabilities, you will predict the frequency of these traits in another class. You will also make a tree diagram for dominant hand, eye color, and hair color that shows the number of different combinations possible.

Procedure

a   Record the frequencies of right-handed people and left-handed people for your group. Record these frequencies in a table similar to the one shown below for an assumed group of five members.

Right-handed   4
Left-handed   1

b   Record the frequencies of different eye colors for your group. Use the four choices of brown, blue, green, and other. Record these frequencies in a table similar to the one shown below.

Brown eyes   3
Blue eyes   1
Green eyes   0
Other   1

c   Record the frequencies of different hair colors for your group. Use the four choices of brown, blonde, red, and black. Record these frequencies in a table similar to the one below.

Brown hair   3
Blonde hair   1
Black hair   1
Red hair   0

d Write each of the three frequency tables for your group on the chalkboard. Combine all your groups' data to make a frequency table for the entire class. The table should show dominant hand, eye color, and hair color frequencies.

e Make a tree diagram showing the different branches for dominant hand, eye color, and hair color. How many branches are in the far-right column? What does this number represent?

f Use the Fundamental Principle of Counting to calculate the number of different combinations of dominant hand, eye color, and hair color. Compare your answer to Step **e**.

g From the frequency table for the entire class, calculate the experimental probability for each trait.

h Based on the experimental probabilities you have calculated, predict how many students in your high school should:

- be left-handed
- have red hair
- have blue eyes

(Your teacher can get the enrollment figures for you.) How can you improve the reliability of your prediction?

i Based on the experimental probabilities you have calculated, determine the probability that a student in your school will be left-handed, *and* have blonde hair, *and* have blue eyes. (You can assume each of these traits is independent for this calculation.)

j How does your experimental probability of being left-handed compare with the commonly accepted probability of 13% for the overall population?

MATH APPLICATIONS

The applications that follow are like the ones you will encounter in many workplaces. Use the mathematics you have learned in this chapter to solve the problems.

Wherever possible, use your calculator to solve the problems that require numerical answers.

Most teachers issue term grades based on the average grade for the period. Some teachers suggest that using the median of a student's grades for the period is a better method. Suppose these are your test grades.

Test Grades:
D	F	D	C
C	B	B	C
B	B	A	B
B			

1 Use the numerical scale of 4 for an A, 3 for a B, 2 for a C, 1 for a D, and 0 for an F. Convert your letter grades to number grades.

2 Determine your mean grade for the period.

3 Determine your median grade for the period.

4 Based on your findings, would you rather have a *mean* teacher or a *median* teacher?

The players on a professional basketball team have the following free-throw percentages at the middle of the season:

Player #	Free-throw %	Player #	Free-throw %
1	73.7	2	86.6
3	85.2	4	86.9
5	71.2	6	75.0
7	77.2	8	85.2
9	86.7	10	73.1
11	99.5	12	80.2

5 Determine the median free-throw percentage for this team.

6 Determine the mean free-throw percentage for this team.

7 A competing team has a higher median free-throw percentage of 85.0% but a lower mean of 77.0%. How can this be possible?

8 Suppose your team trades Player #5 for a player with a free-throw percentage of 79.0%. How will this affect the team's median free-throw percentage? How will it affect its mean free-throw percentage?

You are shopping for an inexpensive room thermometer in a discount store. There are 7 thermometers of the particular style you need on the shelf. However, you notice that they do not seem to agree on the present temperature in the store. Below are the temperatures you read from each of the thermometers.

Temperature (°F)

72 73 74 76
74 73 74

9 Determine the mode of the data.

10 Determine the mean of the data.

11 What is your estimate of the true temperature in the store?

12 Which thermometer would you choose to buy?

A security company has developed a lie detector that analyzes the stress in a person's voice to determine whether the person is telling the truth. The company claims the device to be accurate 94.7% of the time.

13 Suppose that this device is applied to a panel of candidates running for political office. If the claim is true, what is the probability that the device is correct in analyzing the truth of one candidate's statements?

14 Suppose 4 candidates are on the panel. What is the probability that the device can correctly analyze the integrity of all the candidates' statements?

15 What is the probability that the device is *not* able to correctly judge all four candidates' statements?

United States telephone numbers consist of a 3-digit area code, a 3-digit exchange, and a 4-digit station number.

16 In how many ways can you arrange the 3-digit exchange and 4-digit station number to form a telephone number?

17 Suppose that someone randomly dials the area code in a long distance call (in the United States). The exchange and station numbers match yours. What is the probability that the number dialed is your phone number? (Note: Even though some area codes are not valid, such as, 000, it is still possible to dial these numbers.)

During times of national crisis, the U.S. government can draft young men into the armed services. This has been done in the past by randomly assigning numbers from 1 to 366 to the calendar days. Traditionally, two drums of capsules are used. One drum contains number capsules identified 1 to 366. The second drum contains birth-date capsules, one for each of the 366 days of the year (assuming a leap year). One capsule is randomly drawn from each drum so that a number capsule is matched with a date capsule. For example, a number capsule of 128 and a date capsule of April 4 means that those 18-year-olds born on April 4th would be the 128th group to be drafted.

18 What is the theoretical probability that your birth date will be the first date capsule to be drawn?

19 What is the theoretical probability that the number 1 will be the first date capsule drawn from the other drum?

20 What is the probability that your birth date will be drawn first *and* assigned the draft number 1? Express this number as a percent.

21 Consider all the possible draft numbers that can be assigned to your date of birth. What is the probability that your date of birth is assigned the number 1?

22 Why is the probability for Exercise 21 higher than the probability for Exercise 20?

Your company orders a large shipment of scientific calculators from an office supply store. Since the supply house does not have enough of any one brand, it substitutes other brands with similar functions. The packing list reports this distribution of brands.

Brand	Quantity Shipped
A	18
B	30
C	12
D	10

23 If all the calculators are randomly distributed to the company staff, what is the probability that you will receive a Brand A calculator?

24 Still assuming a random distribution of the calculators to the company staff, identify the possible outcomes of such a handout of calculators. Are the possible outcomes resulting from this random distribution of calculators mutually exclusive?

25 Suppose you decide that the Brand B calculator is of equal quality as Brand A. You will be satisfied if you received either. What is the probability that you will receive either Brand A or Brand B?

AGRICULTURE & AGRIBUSINESS

Samples of apples from an orchard were weighed. The weights were rounded to the nearest ounce and recorded.

Sample Apple Weights (oz)

6	6	4
6	7	5
5	8	6
5	5	6
4	5	7
6	5	7
5	5	

26 Construct a tally of the frequencies using one through ten ounces as weight classes.

27 Determine the percent frequency for each class.

28 Construct a percent frequency histogram of the data.

29 What is the mode, median, and mean of the apple weights sampled?

The weights of grain loaded in the bins of a river barge were recorded as follows:

Load	Weight per Bin (ton)
Barley	52.8
Corn	57.0
Flaxseed	49.8
Oats	43.4
Rice	49.1
Rye	52.6
Sorghum	56.6
Soybeans	54.8
Wheat	57.8

30 Compute the mean weight in a bin.

31 Compute the median weight of a bin.

32 Compute the range of weights for these bins.

The table below lists the average wheat yields for several countries during a certain time span.

Average Wheat Yields (pounds per acre)

Brazil	678	Canada	1401
France	2766	India	723
Japan	1999	Mexico	2150
Spain	1008	USA	1552
Russia	946	United Kingdom	3613

33 What is the mean wheat yield?

34 What is the median wheat yield?

35 What is the range of the wheat yields?

You inspect the trees in your orchard for a certain parasite that has appeared in some of your trees. You collect samples from 30 randomly selected trees. Your analysis shows that 12 of the samples are infested and the rest of the samples are clean.

36 What are the two possible events that can occur in your orchard trees with respect to the parasitic infestation?

37 Are these events mutually exclusive? Are they equally likely?

38 What is the experimental probability that a randomly selected tree from your orchard is infested with this parasite?

You read in the newspaper that a man purchased 12 eggs and found every one to be double-yolked. One source stated that the random chance of getting an egg that is double-yolked is 1 in 531.

39 What are the odds of getting a double-yolked egg?

40 If the number of yolks in each of the twelve eggs is independent of one another, what is the probability of getting 12 double-yolked eggs?

41 Suppose your family purchases 12 eggs every week. Approximately how many dozen eggs would your family have to purchase at this rate to expect to find one double-yolked egg inside a carton?

42 What other factors might affect the probability of getting even one double-yolked egg?

BUSINESS & MARKETING

You have asked twenty of your customers to fill out a questionnaire about the quality of your product. One question asks how satisfied they are with their purchase, on a scale of 1 (dissatisfied) to 5 (very satisfied).

Customer Ratings

3	4
1	3
4	4
2	5
4	5
5	4
3	1
4	4
2	3
3	4

43 Make a frequency distribution table for this data.

44 Identify the mode and the median.

45 Compute the mean rating of these customers. Would the mean be a good way to summarize such ratings? Explain your answer.

46 Which measure of central tendency helps you determine if at least half of your customers are reasonably satisfied with their purchase (that is, they give a rating of 4 or 5)?

Your retail toy business frequently experiences short-lived fads. A recent fad involved a particular type of stuffed animal. Your records reflect the rapid rise and decline of sales of this item.

Week #	Items Sold
1	14
2	32
3	68
4	122
5	77
6	39
7	17
8	5
9	2

47 Pick a convenient interval size and draw a histogram to reflect the changing sales for each week.

48 Find the mode for the frequency distribution.

49 Why would the mode of a distribution like this be of interest to a store owner?

50 Suppose the mode occurred later for the next fad. What might the store owner assume about the sales for the weeks that will follow? Is this a valid assumption?

You have recorded the end-of-the-month inventories for your clothing boutique this past year.

End-of-the-month Inventory

18	22	14
20	18	21
28	23	22
17	22	19

51 Determine the mode, median, and mean of this data.

52 Your goal for the end-of-the-month inventory is at least 19 items. How is the median the most useful measure of central tendency in evaluating this goal?

Your company is going to have an open house. A three-member committee is needed to plan the activities. Since all of the workers in your office want to be on the committee, a drawing will be held to elect members. Three names will be selected randomly from a hat containing the names of the seven workers in your office. The first name drawn will be the chairperson of the committee, the second will act as secretary, and the last will be the treasurer.

53 How many different ways are there to construct this committee?

54 Is each of these arrangements of three workers mutually exclusive? Is each equally likely?

55 What is the probability that you will be selected to be chairperson of the committee?

Mortality tables are used by life insurance companies to set insurance premiums. After age 30, mortality rates increase. Thus, insurance companies charge increasingly higher premiums for age groups after 30. The sample mortality table shows the number of persons out of a sample group of 100,000 who are still alive after the indicated age.

Age	Number Living	Age	Number Living
0	100,000	50	089,700
10	098,200	60	082,100
20	097,100	70	065,200
30	096,300	80	039,200
40	094,400		

56 What is the probability that one of these people randomly chosen at birth will live to be 40 years old?

57 What is the probability that one of these people randomly chosen at birth will live to be 80 years old?

58 If one of these people lives to be 60 years old, what is the probability that this person will live another 10 years?

59 If one of these people lives to be 70 years old, what is the probability that this person will live to be 80 years old? Compare this to the answer for Exercise 57.

60 If you were an insurance agent, how would numbers like these influence your decision to accept or reject clients for life insurance, and the premiums that you would ask them to pay?

The drip rate of a patient's IV changes depending on the position of the patient's arm. Over the course of a couple of hours, you count and record the number of drops per minute several times.

Number of Drops per Minute

10	12	10	8
11	11	10	9
11	11		

61 Draw a frequency distribution of the drip rates.

62 What are the values of the mean, mode, and median of this data?

63 What is the range of these data?

A consumer magazine performed a study on the leading brands of black tea, analyzing the caffeine content of a cup of each brand. The table lists the findings, in milligrams of caffeine per cup of brewed tea.

Brand	Caffeine (mg/cup)	Brand	Caffeine (mg/cup)
1	43	14	77
2	39	15	66
3	75	16	51
4	69	17	52
5	10	18	41
6	63	19	44
7	62	20	52
8	61	21	65
9	58	22	74
10	42	23	74
11	69	24	77
12	74	25	44
13	80		

64 Group the data by caffeine content into classes of 10 to 19, 20 to 29, and so on, and draw a histogram of the data.

65 Identify the mode and median of this data.

66 Determine the mean of the data and compare it to the mode and median you determined above.

A person's blood is one of four possible types, A, B, AB, or O, depending on the person's gene pair. The gene pair is the result of the combination of the gene pairs from the person's parents. Type A, for example, results from a combination yielding gene pairs of AA. Type B results from a combination yielding gene pairs of BB. And type AB results from a combination yielding the mixed gene pair of AB. You can chart the combinations of gene pairs for blood type, as you did for flower genes earlier in this chapter.

67 Suppose one parent has a gene pair of AA and the other parent has AB. Make a chart of the possible outcomes of gene pairs for their offspring.

68 List the probabilities for each of the possible blood types.

69 Out of a population of 100 children whose parents have the gene pairs above, approximately how many would you expect to have blood type A? How many would you expect to have blood type AB? How many would have blood type B?

A simple test to determine left-handedness involves spreading apart the ring and middle fingers of each hand to form a "V." A right-handed person can stretch the fingers of his or her left hand farther than the right hand. A left-handed person will have a larger spread on his or her right hand. If there is no difference in the spread, the person tends to be right-handed and has good coordination in both hands. A quick test of this theory found that it was correct in 22 out of 25 cases.

70 A test outcome is correct or incorrect. How many outcomes are possible with the 25 test cases? Are they all equally likely? Does the test of the theory above yield a theoretical probability or an experimental probability?

71 What is the probability of correctly predicting right- or left-handedness when this test is performed on an individual? What is the probability that it is incorrect?

72 You can perform the test with yourself and your classmates. Do your results agree with the results given above?

FAMILY AND CONSUMER SCIENCE

You are surveying the prices for carpet padding for a client. You record the following prices from several different suppliers:

Carpet Padding Price ($ per sq. yd)

$2.09	$3.55
$2.75	$3.15
$2.99	$3.07
$2.95	

73 What is the range in carpet padding prices?

74 What is the mean carpet padding price?

75 What is the median carpet padding price?

You have recorded the temperatures from a refrigerated dairy case during a six-day period, twice each day.

Temperature (°F)		
Day 1	38	39
Day 2	38	40
Day 3	40	39
Day 4	40	38
Day 5	38	39
Day 6	38	40

76 What is the range of the temperatures?

77 What is the mean temperature?

78 What is the median temperature?

79 What is the mode of the temperature data?

Suppose that you have five different wall hangings and three locations on which to place them.

80 How many different ways can you arrange the five wall hangings?

81 If you found one more wall hanging to include in the arrangement, what would be the new number of choices available to you?

A large bowl is filled with a nut mix. Since your favorite type of nut is the almond, you decide to study the number of almonds in the bowl. You remove a few handfuls of nuts, then count and record the types.

| Cashews | 5 | Peanuts | 54 |
| Brazil nuts | 2 | Almonds | 7 |

82 Determine the probability of randomly removing each type of nut in the mix.

83 Suppose you remove one nut at random. You return the nut and remove another one. What is the probability of obtaining an almond on the second draw?

84 Suppose you randomly remove two nuts from the bowl. What is the probability of obtaining an almond?

INDUSTRIAL TECHNOLOGY

You find the height of a stack of twenty pieces of plywood to be $7\frac{5}{8}$ inches.

85 What is the mean thickness of the pieces of plywood in this stack?

86 If you knew the boards were supposed to be either $\frac{1}{4}$, $\frac{3}{8}$, $\frac{1}{2}$, or $\frac{5}{8}$ inches thick, which thickness would you suspect the boards to be?

While overhauling an eight-cylinder engine, you forget to number the eight rod caps. When you reassemble the rods, you do not know which cap to match with each rod.

87 How many combinations of rods and caps are possible?

88 What is the probability of "accidentally" matching all eight caps with the correct rods on the first try?

A certain product is assembled in stages using three machines: Machine A, Machine B, and Machine C. During the previous six months, each of the machines has broken down, causing a slowdown in production while it was being repaired. You have kept a record of the last six months' history.

Machine	Days Operating	Days Down for Repair
A	116	16
B	106	26
C	122	10

89 For each machine, determine the experimental probability of the machine being down on any given day.

90 If all machines are down at the same time, production must cease. Assuming that each machine's status is independent of the others, determine the probability that all three machines will be down at the same time.

A factory assembles deodorant bottles. The bottles are randomly selected from the lot and assembled with randomly selected bottle caps. A quality engineering study estimates that about 3% of a lot of 1000 bottles were made with tops of too small a diameter. Unfortunately, these bottles were matched up with a lot of 1000 bottle caps, of which 1.8% were indicated to have too large a diameter. Whenever bottle tops that are too small are combined with caps that are too large, the bottles will leak.

91 What is the probability that a small bottle top in this lot is matched up with a large bottle cap?

92 How many of the lot of 1000 bottles should you expect to leak because of this mismatch of bottle tops and caps?

Your company has two safety vehicles; each is operable 95% of the time. Assume that the availability of each vehicle is independent of the other's condition.

93 What is the probability that both vehicles are operable at the same time?

94 What is the probability that only one of the two vehicles is operable at any time?

95 What is the probability that neither vehicle is operable at the same time?

A company produces a certain type of candy. Each bag contains small candies that can have one of five different color coatings: red, orange, brown, green, and yellow. Each bag, on the average, contains about 45 candies. The bags are filled from a bin that has a uniform mixture of each of these five colors.

96 What is the probability that the first piece of candy into a bag is red?

97 What is the probability that the first piece of candy into a bag is not red?

98 What is the probability that the second piece of candy into a bag is not red?

99 What is the probability that both the first *and* second candies into the bag are not red?

100 What is the probability that a bag gets filled with 45 pieces of candy from the bin, and not one of the pieces is red?

CHAPTER 7   ASSESSMENT

Applications

The highest annual commissions paid on sales by the Top-Flite Shoe Corporation were listed in its newsletter.

Region	Commission
1	$10,000
2	7,500
3	9,000
4	8,000
5	10,000
6	9,500

1. What is the mode of the commissions?

2. What is the median of the commissions?

3. What is the mean of the commissions?

4. Which measure of central tendency would you report to attract new salespeople interested in making a large commission?

5. Suppose the person in Region 3 is replaced by a salesperson who makes $6000 more in commissions. How is the mean affected?

The following final exam grades were posted in the Library:

98	94	94	74	87
100	80	89	96	68
88	88	98	92	79
67	97	94	84	76

6. Make a frequency table using the following classes:

61–70   71–80   81–90   91–100

7. What is the mean of the test scores?

8. Use the frequency distribution to create a histogram.

Use the following data:

(7, 7)   (3, 5)   (9, 6)   (1, 2)   (5, 3)

9. Make a scatter plot.

10. Describe the correlation exhibited by the scatter plot.

A six-sided, colored cube has two faces painted red, two faces painted blue, one face painted green, and one face painted yellow.

11. If the cube is rolled once, what is the probability it will come up red?

12. If the cube is rolled once, what is the probability it will come up blue or green?

13. If the cube is rolled twice, what is the probability the first roll will be green and the second roll yellow?

A license plate has seven positions. You can place any letter in the first position and any digit in the other positions.

14. How many arrangements of the letters and digits are possible?

15. What is the probability of randomly choosing a plate that begins with the letter A?

Math Labs

16. You hold a meterstick at the 100 centimeter end and drop it for Ann and John. Ann catches the stick at the 70 centimeter mark. John catches the stick at the 82 centimeter mark. Who has the quickest reaction time?

17. The histograms for the single-die throws were relatively flat. However, for the two-dice throws, the histograms have a higher frequency in the middle than at the ends. Explain why this is so.

18. Construct a tree diagram for left- or right-handed; dark brown, light brown, blue, green, or hazel eyes; and black, brown, blonde, or red hair. How many combinations are possible?

CHAPTER 8

In today's highly competitive world, businesses are constantly looking for solutions that save money or conserve scarce resources. Skillful problem solvers use such tools as systems of equations to find optimum solutions. You can take an early step toward career success by mastering the algebra skills found in this chapter.

Contents: Systems of Equations

Systems of Equations

Objectives

1. Solve a system of two equations by graphing.
2. Solve a system of two equations by substitution.
3. Solve a system of two equations by addition or subtraction.
4. Solve a system of two equations by using determinants.
5. Use a system of equations to solve problems involving two unknowns.

In Chapters four through six, you used equations with one variable to model problems involving volume, speed, and money. However, some situations involve two variables and need two equations to model the problem. In these cases, you must set up **systems of equations** to solve the problems.

For instance, when a large aerospace company begins planning a new airplane, its engineers and salespeople use systems of equations to model production costs and projected sales revenue. This model can predict whether the new plane will earn a profit or cause the company to lose money.

In this chapter, you will examine several ways to solve a system of equations. These ways will include:

- *Using* graphing *to find points of intersection.*
- *Using* substitution *to reduce two equations with two unknowns to one equation with one unknown.*
- *Using* addition *or* subtraction *to reduce two equations with two unknowns to one equation with one unknown.*
- *Using* determinants *to find the unknowns in systems of equations.*

As you watch the video for this chapter, look for ways in which systems of equations are used in the workplace.

LESSON 8.1 SOLVING A LINEAR SYSTEM BY GRAPHING

Mr. Cassidy is trying to choose the best one-day rental plan from the A-1 Rental Agency.

The agency offers two rental plans:

Plan 1: $30.00 per day with unlimited mileage
Plan 2: $20.00 per day plus 10 cents per mile

To compare the two plans, you can model each with a function. Let C_1 represent the cost under Plan 1 and C_2 represent the cost under Plan 2. Let N represent the number of miles driven.

Plan 1:   Cost = $30.00

$$C_1 = 30$$

Plan 2:   Cost = ($0.10 per mile) ($N$ miles) + $20.00

$$C_2 = 0.10N + 20$$

Plan 1 is a constant function. Plan 2 is a linear function with slope 0.10 and y-intercept 20. When you graph the two functions on the same coordinate axes, you can see the solution.

N	0	50	100	150
C_1	30	30	30	30

N	0	50	100	150
C_2	20	25	30	35

At the the point where the two graphs intersect, the plans cost the same. This is called the **break-even point.** The graph shows the break-even point occurs when the x-coordinate is 100. In other words,

- *Both plans cost the same if Mr. Cassidy drives exactly 100 miles.*
- *Plan 1 is less expensive if he drives over 100 miles.*
- *Plan 2 is less expensive if he drives under 100 miles.*

The two functions modeling the car-rental problem represent a **system of linear equations** in two variables. The following is another example of such a system:

$$y = 3x - 5$$
$$y = -x + 7$$

The solution to each linear equation or function is a set of ordered pairs. To solve the *system,* you must find all of the ordered pairs that solve *both* equations. These ordered pairs represent the points of intersection of the graphs of the system of equations

Examine the table of values for each equation.

Only the ordered pair (3,4) solves both equations. Thus, the ordered pair (3,4) is the solution to this system of equations.

Sometimes you cannot find the solution by examining a table of

$y = 3x - 5$		$y = -x + 7$	
x	y	x	y
−1	−8	−1	8
0	−5	0	7
1	−2	1	6
2	1	2	5
3	4	3	4
4	7	4	3
5	10	5	2
⋮	⋮	⋮	⋮

values. However, you can graph the equations. The solution is the ordered pair representing the point where the graphs intersect.

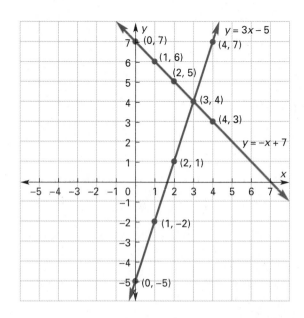

If you use a graphics calculator, you must rewrite the equations in slope-intercept form. When using graph paper, it is often easier to use the standard form of the equation and draw a line between the x-intercept and y-intercept. In either case, check your solution and make sure it is reasonable.

EXAMPLE 1 The Point of Intersection

Use graph paper or a graphics calculator to solve this system.

$$x - y = 3$$
$$2x + y = 3$$

SOLUTION

If you are using a graphics calculator, rewrite each equation in slope-intercept form, $y = mx + b$.

$$y = x - 3 \qquad\qquad y = -2x + 3$$

x	y
-1	-4
0	-3
1	-2
2	-1
3	0

x	y
-1	6
0	3
1	1
2	-1
3	-3

Graph the system and find the point of intersection.

The solution is $(2, -1)$.

To check the solution, substitute 2 for x and -1 for y in the original equations.

$$x - y = 3 \qquad\qquad 2x + y = 3$$

$$2 - (-1) = 3 \qquad (2)(2) + (-1) = 3$$

$$3 = 3 \qquad\qquad 3 = 3$$

EXAMPLE 2 Solving a System Graphically

The band director must order 35 uniforms. There are usually five more boys than twice the number of girls in the band. How many uniforms of each type should the band director order?

SOLUTION

Write a system of equations to model the situation. Then solve the system using graph paper or a graphics calculator.

Let x represent the number of girls.

Let y represent the number of boys.

Since there are 35 band members, $x + y = 35$ or $y = 35 - x$.

The number of boys is five more than two times the number of girls. Thus,

$$y = 2x + 5$$

Now solve the system.

There are 10 girls and 25 boys. Check the solution in the original equations to make sure it is correct.

When you use a graphics calculator, make sure you use a friendly window. You may also need to use TRACE and ZOOM to find the point of intersection.

ACTIVITY | **Using a Graphics Calculator to Solve a System**

1 Graph $y = 4x - 5$ and $y = -x + 7$ on your graphics calculator.

2 Use TRACE to find the intersection of the graphs.

3 Write the coordinates of the intersection.

LESSON ASSESSMENT

Think and Discuss

1 How can you use the graphs of two equations to find the solution to a system of equations?

2 In what form must you write an equation to enter it in a graphics calculator?

3 If two straight lines intersect on a graph, how many points of intersection are there?

Practice and Problem Solving

Find the solution to each system graphically.

4. $y = 2x - 7$
$y = -3x + 3$

5. $y = -x + 7$
$y = 2x - 5$

6. $y = 2x - 4$
$y = 3x - 7$

7. $y = -2x + 3$
$y = 5x + 10$

8. $3x + y = 6$
$x + 2y = 2$

9. $3y - 2x = 12$
$2y + x = 8$

10. $2x + y = 9$
$x + y = 4$

11. $5x - 2y = 18$
$x + 2y = -6$

12. $4x + 2y = -2$
$3x + y = -2$

Write a system of equations to model each situation. Solve the system graphically.

13. In a basketball game, Juan made a total of twelve 2-point and 3-point baskets totaling 26 points. How many 2-point shots and how many 3-point shots did Juan make?

14. A class divides into two groups for a science experiment. The first group has five fewer students than the second group. If there are 27 students in the class, how many students are in each group?

Let $f(x) = |x| + 3$.

15. Graph f.

16. What is the domain of f?

17. What is the range of f?

18. What is $f(-3)$?

Latisha recorded the number of hours she worked during a 2-week period in July.

6, 6, 7, 8, 7, 6, 4, 7, 5, 8

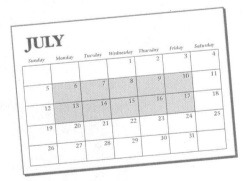

19. List the data in order from least to greatest.

20. Find the mode of the data.

21. Find the median of the data.

22. Find the mean of the data.

LESSON 8.2 SPECIAL LINEAR SYSTEMS

When you graph two equations on the same coordinate axes, one of three things can happen.

The lines may intersect. The system has one solution.

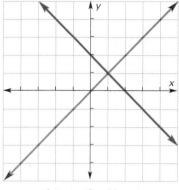

Intersecting Lines

The lines may be parallel. The system has *no* solution.

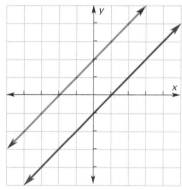

Parallel Lines

The lines may coincide. The system has infinite solutions.

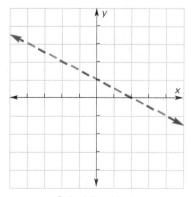

Coinciding Lines

In the worlds of agriculture, business, engineering, industry, and science, systems of linear equations model many problems. Most of these linear systems have single, or unique solutions. A system that has at least one solution is called **consistent**. If the system has *only* one solution, the system is **independent**.

ACTIVITY 1 Modeling Profit in a Business

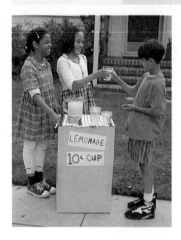

Your younger sister wants to earn money by selling lemonade. The cost of starting the business is $1.20, and each cup of lemonade costs six cents to make.

Your sister's expenses in dollars (d) is equal to six cents times the number of cups sold (n) plus the $1.20 start-up cost.

$$d = 0.06n + 1.20$$

If the lemonade sells for ten cents a cup, the income (d) is equal to ten cents times the number (n) of cups sold.

$$d = 0.10n$$

1 How does this graph model your sister's profit or loss?

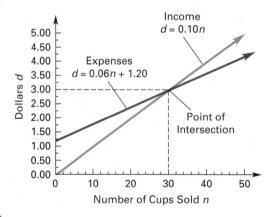

2 What does the point (0, 0) mean on the income graph?

3 What are some of the start-up costs?

4 If she sells 20 cups of lemonade, what is her profit or loss?

5 If she sells 40 cups, what is her profit or loss?

6 What are the coordinates for the break-even point?

7 How many cups must she sell to make a profit?

8 Is this system of equations consistent? Is it independent?

ACTIVITY 2   The Profit Line Changes

Lemons
$1.⁰⁰ / lb

Suppose the cost of making lemonade increases to 10 cents per cup. The equation that models the expenses becomes

$$d = 0.10n + 1.20$$

1 Graph the system $d = 0.10n + 1.20$
$$d = 0.10n$$

2 What is the slope of each line?

3 What is the y-intercept of each line?

4 How are the lines related?

5 What is her profit or loss if she sells 20 cups? 40 cups?

6 What is the break-even point?

7 Is this system consistent?

The lines in Activity 2 will never intersect. The expression $0.10n + 1.20$ will never equal $0.10n$. The two lines are **parallel.** If two lines are parallel, they have the same slopes but different y-intercepts.

> **Parallel Lines**
> Nonvertical lines are parallel if they have the same slope.
>
> Vertical lines are also parallel even though their slopes are undefined.

Since the lines in Activity 2 never intersect, the system has *no solution.* A system with no solution is called **inconsistent.**

EXAMPLE An Inconsistent System

Solve this system.

$$-2x + y = -1$$
$$-6x + 3y = 12$$

Use graph paper or a graphics calculator. First write each equation in slope-intercept form.

$$y = 2x - 1$$
$$y = 2x + 4$$

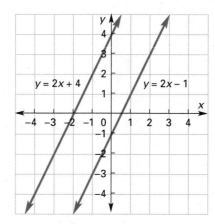

Notice that the slopes are equal, but the y-intercepts are different. Therefore, the lines are parallel. There is no solution to this system, and it is inconsistent.

Ongoing Assessment

Which of the systems is inconsistent?

a. $x + y = 5$ **b.** $x + y = 5$
 $x - y = 3$ $x + y = 4$

ACTIVITY 3 **A Dependent System**

1 Graph this system on the same coordinate axis.

$$4x + y = -2$$
$$8x + 2y = -4$$

2 What is the slope of each line?

3 What is the y-intercept of each line?

4 What is the intersection of the two lines?

5 Is this system consistent? Is it independent? How many solutions does the system have?

In Activity 3, the lines are the same. When each equation is written in slope-intercept form, the equations are identical. Every solution for one equation is also a solution for the other equation. The system is consistent because the system has at least one solution. In fact, the system has infinite solutions. The system is **dependent** because all the solutions are the same.

Critical Thinking   Explain the difference between the following three systems:

a. a system that is consistent and dependent
b. a system that is consistent and independent
c. a system that is inconsistent

LESSON ASSESSMENT

Think and Discuss

1 Describe the graphs and slopes of a system of equations with no solution.

2 Describe the graphs, slopes, and y-intercepts of a system of equations with an infinite number of solutions.

3 Describe the graphs of a system of equations that has one solution.

4 What is the difference between consistent and inconsistent systems of equations?

5 What do you mean when you say a system of equations is dependent or independent?

Practice and Problem Solving

For Exercises 6-17, state whether the system is

a. consistent and independent
b. consistent and dependent
c. inconsistent

6.   **7.**   **8.**

9.   **10.**   **11.**

12.   **13.**   **14.**

15. $2x = y + 3$

$2y = 4x + 5$

16. $2x = 10y + 4$

$x - 5y = 2$

17. $x + y = 5$

$3x + 3y = 21$

Write a system of linear equations to model the number of flights each flight attendant works over two consecutive weeks. Identify each system as independent, dependent, or inconsistent.

18. Keesha works a total of 22 flights over two weeks. Her average number of flights over the two weeks is 11.

19. Sal works a total of 22 flights over two weeks. The second week, he works four fewer flights than the first week.

20. Heather works a total of 22 flights over two weeks. The second week, she works eleven minus the number of flights she worked the first week.

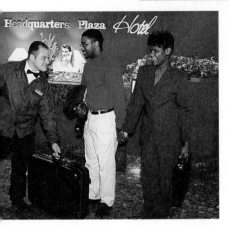

The following table shows the number of miles a person of each occupation walks in one day.

A. City messenger	4.0	B. Dentist	0.8
C. Hospital nurse	5.3	D. Hotel employee	3.2
E. Secretary	2.2	F. Teacher	1.7

21. If two miles is the average distance walked in one day on the job for people of all occupations, use an integer to indicate the deviation from the average for each occupation listed in the table.

22. Draw a number line. Place the deviation from the average for each occupation (A–F) on the number line.

Let r represent the radius of the Earth. The formula for the circumference (C) of the Earth is $C = 2\pi r$. The formula for the surface area (S) of the Earth is $S = 4\pi r^2$. The average radius of the Earth is 6.37×10^6 kilometers at the equator.

23. If a road could be built around the Earth at the equator, how long would the road be?

24. If a road could be built on an elevated bridge around the equator just 1 meter above the Earth, what would be the length of the road?

25. What is the surface area of the Earth?

26. If approximately 70% of the Earth is covered by water, how many square kilometers of the Earth's area is water?

27. Suppose the radius of the Earth is doubled. By what factor is the circumference increased? By what factor is the surface area increased?

LESSON 8.3 SOLVING SYSTEMS BY SUBSTITUTION

Once again, recall Mr. Cassidy's car rental problem. You can use another method to choose the best plan.

Plan 1:   Cost = $30.00

$$C_1 = 30$$

Plan 2:   Cost = ($0.10 per mile) ($N$ miles) + $20.00

$$C_2 = 0.10\,N + 20$$

For what number of miles driven is the cost the same for both plans? In an earlier lesson, you used a graph to find that the break-even point occurs when N is 100 miles. Another way to find the break-even point uses an algebraic technique called **substitution**.

The break-even point is the value of N that makes C_1 and C_2 the same. Since $C_1 = C_2$, you can substitute the value of C_1 (30) for C_2 in the second equation and solve for N.

Thus, when $C_1 = C_2$, the following substitution is possible.

$$30 = 0.10N + 20$$

$$0.10N = 10$$

$$N = 100$$

You can use the method of substitution to solve any system. For example, consider this system.

$$3x - y = 1$$
$$2x + y = 9$$

To solve this system by substitution, follow these steps.

1. Solve the first equation, $3x - y = 1$, for y.

$$y = 3x - 1$$

2. Substitute the expression $3x - 1$ for y in the second equation.

$$2x + (3x - 1) = 9$$

3. Solve the equation in Step 2 for x.

$$5x - 1 = 9$$
$$5x = 10$$
$$x = 2$$

4. Substitute 2 for the value of x in the first equation and solve for y.

$$3x - y = 1$$
$$3(2) - y = 1$$
$$y = 6 - 1$$
$$y = 5$$

5. You have now found a solution $(2, 5)$ for the system. Check your solution in each of the original equations.

$$3x - y = 1 \qquad 2x + y = 9$$

$$3(2) - 5 = 1 \qquad 2(2) + 5 = 9$$

$$6 - 5 = 1 \qquad 4 + 5 = 9$$

The solution is correct. As a further check, use graph paper or a graphics calculator to graph the system.

ACTIVITY 1   Special Systems

1 Solve the following system by substitution.

$$-2x + y = -1$$
$$-6x + 3y = 12$$

2 Is the system consistent or inconsistent?

3 What happens to the variables when you solve an inconsistent system by substitution? Does the equation simplify to a true statement?

4 What are the solutions to the system?

5 Solve the following system by substitution.

$$4x + y = -2$$
$$8x + 2y = -4$$

6 Is the system dependent or independent?

7 What happens to the variables when you solve a dependent system by substitution? Does the equation simplify to a true statement?

8 What is the solution for the system?

EXAMPLE The Perimeter of a Rectangle

Suppose the perimeter of a rectangular picture frame is 56 inches. If the width is 6 inches longer than the length, what is the area enclosed by the picture frame?

SOLUTION

Let h represent the height of the frame. Let w represent the width of the frame.

The perimeter is $2h + 2w = 56$.

The width is the height plus six, or $w = h + 6$.

Solve the system $2h + 2w = 56$ by substitution.

$$w = h + 6$$

Substitute $h + 6$ for w in the first equation and solve for h.

$$2h + 2(h + 6) = 56$$

$$2h + 2h + 12 = 56$$

$$4h = 44$$

$$h = 11$$

Substitute 11 for h in the equation $w = h + 6$ and solve for w.

$$w = 11 + 6 = 17$$

Check the solution (11, 17) in each of the original equations.

$$2h + 2w = 56 \qquad w = h + 6$$

$$2(11) + 2(17) = 56 \qquad 17 = 11 + 6$$

$$22 + 34 = 56$$

The height of the frame is 11 inches, and the width is 17 inches. Thus, the area enclosed by the frame is $(11)(17) = 187$ or 187 square inches.

The first step in solving a system by substitution is to isolate one of the variables in one of the equations. You can always solve the system regardless of which variable you choose to isolate. But sometimes one of the variables is easier to isolate than the other.

ACTIVITY 2   Solving for Either Variable

Consider this system.

$$a - 4b = -8$$
$$3a + 5b = 27$$

1. Solve the first equation for a and then for b.

2. Does solving for a or b result in the simplest substitution?

3. Substitute the simplest expression into Equation 2 and solve the system.

Now rework the problem beginning with Equation 2.

4. Solve the second equation for a and then for b.

5. Does solving for a or for b result in the simplest substitution?

6. Substitute the simplest expression into the first equation and solve the system.

7. Check your solution using a graphics calculator.

1. Plot the data points for plasma source voltage as a function of average thickness of gold deposited for a target thickness of 8 microns. Use your graph to estimate the plasma source voltage for an 8 micron average deposition.

2. For a more accurate value, pick two data points for voltage (V) and two for thickness (t) for a target thickness of 8 microns. Using these data points, write two equations in the form $V = mt + b$. Solve these two equations for the slope and intercept of the line. Then calculate the plasma source voltage for an 8 micron deposition.

Date; Time of Production	Target Thickness of Gold Deposition (microns)	Average Thickness of Gold Deposited (microns)	Plasma Source Voltage (kilovolts)	Time of Exposure (hours)	Average Temperature (Celsius)
03/12;AM	8.0	15.1	20.1	4.67	192
03/12;PM	8.0	12.6	18.1	4.35	195
03/13;AM	8.0	4.2	9.1	4.14	189
03/13;PM	8.0	4.9	10.1	4.33	
03/14;AM	12.0	6.4			
03/14;PM	12.0				
03/15;AM	12.0				
03/15;PM	12.0				

Susan:

This is a portion of my team's data log from last week's solar cell production runs. As you can see, we are having a hard time controlling the thickness of gold in the plasma deposition process. Our specs called for target thicknesses of 8 and 12 microns, and we were off target every run.

Tomorrow we start another production run, and we need better control over the gold thickness. Can you adjust the plasma source voltage for an 8 micron deposition?

Thanks,
Anthony

LESSON ASSESSMENT

Think and Discuss

1 Describe the steps for solving a system of equations by substitution.

2 Given a system of equations that is inconsistent, what happens when you solve the system by substitution? Give an example.

3 In a system of equations that is consistent and dependent, what happens when you solve the system by substitution? Give an example.

Solve by substitution. Identify the system as inconsistent, dependent, or independent.

4. $x = -2y + 4$
$3x - y = 26$

5. $2x + 4y = 12$
$2y = 6 - x$

6. $x + y = 0$
$y = x - 2$

7. $x - 5y = 6$
$2x + 3y = 25$

8. $5x + 3y = 10$
$2x + y = 3$

9. $x - y = 8$
$3x + 3y = 24$

10. $6x + 2y = 5$
$3x + y = 3$

11. $4x + 3y = 6$
$2x + y = 4$

12. $2x + y = -1$
$4x + 2y = -4$

Write a system of equations to model each situation. Solve the system by substitution. Check the solution.

13. At the beginning of school, Nancy bought one 3-subject notebook and four single-subject notebooks for $15. If the 3-subject notebook cost twice as much as a single-subject notebook, how much did each type of notebook cost?

14. Artier has a total of $7000 invested, part in a mutual fund and part in a certificate of deposit (CD). The CD pays 7% interest, and the mutual fund pays 9%. If he receives $550 per year in interest, how much does Artier have in each investment?

15. Together Nita and Joe do 95 haircuts a week. If Nita does 16 fewer than twice as many as Joe, how many haircuts does each person do?

The sum (*S*) of the measures of the angles of a polygon with *n* sides is found by using the formula

$$S = 180(n - 2)$$

Find the total number of degrees in each polygon.

16. A triangle.   **17.** A quadrilateral.   **18.** A hexagon.

19. The measures of the angles in a triangle are represented by *a*, *a* + 20, and *a* + 25. Find the measure of each angle.

20. The measures of the angles in a parallelogram are represented by 2*m*, 2*m*, 3*m* − 10, and 3*m* − 10. Find the measure of each angle.

Suppose you have a hat containing 10 tickets. The tickets are numbered 1, 2, and so on through 10. You draw one ticket from the hat. What is the probability the number on the ticket will be

21. Even?   **22.** A prime number?

23. Greater than 3?   **24.** Odd and less than 8?

You remove one ticket and look at it. You put the ticket back in the hat and draw again. What is the probability that

25. The number on the first ticket is even, and the number on the second ticket is odd?

26. The number on the first ticket is odd, or the number on the second ticket is odd?

27. The sum of the numbers on the two tickets is 10?

28. The sum of the numbers on the two tickets is greater than 18?

LESSON 8.4 SOLVING SYSTEMS BY ADDITION OR SUBTRACTION

Nancy and Jim are team leaders at the Allied Computer Chip Manufacturing Company. The production supervisor needs to report the number of computer chips each team made on Friday. The supervisor knows the total number of chips produced by both teams is 130. Nancy's team made 10 more chips than Jim's team.

Let x represent the number of chips made by Nancy's team. Let y represent the number of chips made by Jim's team.

The following system models the information known by the supervisor.

$$x + y = 130$$
$$x - y = 10$$

Both the graphing and substitution methods will solve this system. However, you can also use another method based on an alternate form of the Addition Property of Equality.

Addition Property of Equality (Alternate Form)

For all numbers a, b, c, and d,

if $a = b$ and $c = d$, then $a + c = b + d$.

This property allows you to form a new equation from two existing equations. For example, suppose you have the following system of equations:

$$a = b$$
$$c = d$$

If you add the left sides and then the right sides, a new equation results.

$$
\begin{array}{r}
a = b \\
+\ c = d \\
\hline
a + c = b + d
\end{array}
$$

The reason for using the Addition Property of Equality to form a new equation is that in certain cases the new equation will contain variables that are opposites. Since opposites sum to zero, the new equation will contain only one variable, which you can solve using the methods described in Chapters 4–6. For instance, start with the system that models computer chip production.

$$x + y = 130$$
$$x - y = 10$$

Now use the Addition Property of Equality to add the left and right sides of these two equations.

$$(x + y) + (x - y) = 130 + 10$$

Notice that y and $-y$ are opposites, and their sum is zero. Thus the y variable is eliminated, leaving an equation with only the single variable x.

$$(x + x) + 0 = 130 + 10$$
$$2x = 140$$

Chip Production

Nancy's Team	Jim's Team
70	60

The solution to this equation is $x = 70$. Substituting this value for x in either of the original equations results in $y = 60$. Since the ordered pair $(70, 60)$ is the solution for the system, Nancy's team made 70 computer chips and Jim's team made 60 computer chips on Friday.

Ongoing Assessment

Solve this system using the alternate form of the Addition Property of Equality.

$$x + y = 100$$
$$x - y = 20$$

You can also use an alternate form of the Subtraction Property of Equality to solve systems of linear equations.

Subtraction Property of Equality (Alternate Form)

For all numbers a, b, c, and d,

If $a = b$ and $c = d$, then $a - c = b - d$.

Subtraction can also solve the computer chip production equations.

$$x + y = 130$$
$$x - y = 10$$

This time subtract the expressions on each side of the equal signs. Notice in the following steps how the x-variable is eliminated.

$$x + y = 130 \qquad \Rightarrow \qquad x + y = 130$$
$$-(x - y) = -10 \qquad \Rightarrow \qquad -x + y = -10$$

$$2y = 120$$
$$y = 60$$

The result of replacing y by 60 in either original equation gives $x = 70$.

EXAMPLE

Solve this system.

$$2x + 2y = 10$$
$$3x + 2y = 14$$

SOLUTION

Use the alternate form of the Subtraction Property of Equality to subtract the expressions on the left side and the expressions on the right side of the equal sign.

$$2x + 2y = 10 \qquad \Rightarrow \qquad 2x + 2y = 10$$
$$-(3x + 2y) = -14 \qquad \Rightarrow \qquad -3x - 2y = -14$$

$$-x + 0y = -4$$
$$-x = -4$$
$$x = 4$$

To solve for y, substitute 4 for x in the first equation.

$$2(4) + 2y = 10$$
$$8 + 2y = 10$$
$$2y = 2$$
$$y = 1$$

The solution to the system is (4, 1). Check this solution in both equations.

Use the alternate form of the Subtraction Property of Equality to solve the following system:

$$2x + y = -2$$
$$2x + 3y = 0$$

ACTIVITY   **Solving a System by Subtracting**

On Monday, two computer chip manufacturing teams produced a total of 80 chips. Nancy's team made four fewer than twice the number produced by Jim's team.

Let x represent the number of items made by Nancy's team. Let y represent the number of items made by Jim's team.

1 What equation models the total number of chips made by the teams?

2 Write an equation that models the number of chips made by Nancy's team.

3 Write the equations from Steps 1 and 2 in the standard form $ax + bx = c$.

4 Solve this system using the alternate form of the Subtraction Property of Equality.

5 How many chips did each team produce on Monday?

You can use the addition or subtraction method to solve a linear system. Write each equation in standard form, $ax + by = c$. Add or subtract the terms on the same side of the equal signs of the two equations to eliminate one of the variables from the system. Then solve for the other variable.

Solve this system by the addition or subtraction method.

$$2x = 3y - 5$$
$$3y + 4x = 8$$

CULTURAL CONNECTION

Chinese mathematicians solved systems of equations even before they invented the abacus. These mathematicians used colored rods to represent numbers in the system. For example, let green rods represent positive numbers and red rods represent negative numbers.

Explain how the diagram is used to solve the system for x.

$$3x + y = 3$$

$$-x - y = 1$$

Use the Chinese method to solve for y. What is the solution to the system?

LESSON ASSESSMENT

Think and Discuss

1 When is the addition method the best method for solving a system of equations in standard form? When is the subtraction method best?

2 How do you write an equation in standard form?

3 Why should the two equations be written in standard form before a system is solved?

Solve by addition or subtraction.

4. $2y - 2x = 3$
$2y - 5x = 9$

5. $5y + 3x = 0$
$3y - 3x = 0$

6. $6x - 4y = 4$
$4y - 6x = 2$

7. $x - 3y = -6$
$x + y = 6$

8. $x + y = 2$
$2x - y = 1$

9. $3x + y = 6.5$
$x + y = 2.5$

10. $2x + y = 2$
$2x + 3y = 14$

11. $x - y = 2$
$x + y = 4$

12. $x - y = -2$
$2x - y = 1$

Write a system of equations for each situation. Solve and check your answers.

13. Because of limited storage space at the job site, a bricklayer must schedule two deliveries. The two deliveries must contain a total of 23 pallets of bricks. One delivery must have three more pallets than the other. How many pallets of bricks will be in each delivery?

14. In a surveyor's drawing using a coordinate system, one side of an angle lies along the line $x + y = 3$, and the other side lies along the line $2x + y = -3$. Find the coordinates of the vertex of the angle.

15. To send a telegram, you must pay a fixed (or flat) rate for the first ten words and a fixed charge for each additional word. The cost to send a 17-word telegram is $14.55. The cost for a 21-word telegram is $17.15. What is the flat rate, and what is the fixed charge for each additional word?

The American Sand and Gravel Company charges $215 for one ton of sand. One of the American Sand and Gravel Company trucks can haul a load of sand in a bed that measures 8 feet by 4 feet by 2 feet. One cubic yard of sand weighs about 1.212 tons.

16. How many cubic yards of sand will one truck hold?

17. How many tons of sand will one truck hold?

18. If one driver makes six deliveries to a construction site, what is the cost of the sand delivered?

19. Write a formula to find the number of truckloads (N) that can be delivered for x dollars.

Let $g(x) = \dfrac{1}{x}$.

20. Find $g(5)$.

21. Find $g(0.5)$.

22. Graph g in intervals of 0.5 for the domain from 0.5 through 5.

23. What is the range for the domain given in Exercise 22?

24. What happens to $g(x)$ as x becomes very large?

25. What happens to $g(x)$ as x becomes very small?

LESSON 8.5 SOLVING SYSTEMS BY MULTIPLICATION

In the last lesson, you used addition and subtraction to solve systems of linear equations. This resulted in the elimination of one of the variables. However, most of the time the variables are not opposites, and another step is needed.

EXAMPLE 1 Use Multiplication First

Solve this system.

$$2x = 4y - 5$$
$$3y + 4x = 12$$

SOLUTION

First write the equations in standard form.

$$2x - 4y = -5$$
$$4x + 3y = 12$$

None of the terms are opposites. Thus, adding or subracting will not eliminate a variable. However, if you use the Multiplication Property of Equality to multiply both sides of the first equation by -2, opposite terms will result. The sum of these opposite terms will equal zero, thus eliminating the x-variable.

$$
\begin{array}{lll}
-2(2x - 4y) = -2(-5) & \Rightarrow & -4x + 8y = 10 \\
4x + 3y = 12 & \Rightarrow & +\ \underline{4x + 3y = 12} \\
& & 0 + 11y = 22 \\
& & y = 2
\end{array}
$$

Substitute 2 for y in either of the original equations to find $x = 1.5$.

Check to make sure $(1.5, 2)$ is the solution to the system.

Ongoing Assessment

Solve this system of equations using multiplication and then addition.

$$3x - 5 = y$$
$$5 = x - 2y$$

Rachel works as a chemist's assistant. For an experiment, Rachel needs 25 gallons of a 24% alcohol solution. She has two solutions available: a 20% alcohol solution and a 30% alcohol solution. How much of each solution should Rachel mix to get a 24% alcohol solution?

You can model this problem using a system of linear equations.

Let x represent the number of gallons of 20% solution.

Let y represent the number of gallons of 30% solution.

The total number of gallons of the 24% solution is

$$x + y = 25$$

The alcohol in the 20% solution (20% of x gallons) plus the alcohol in the 30% solution (30% of y gallons) must equal the total alcohol in the 24% solution (24% of 25 gallons). That is,

$$20\% \, x + 30\% \, y = 24\%(25) \text{ or}$$
$$0.20x + 0.30y = 6$$

Rachel must solve this system.

$$x + y = 25$$
$$0.20x + 0.30y = 6$$

First, multiply each side of the second equation by 100 to eliminate the decimals.

$$x + y = 25$$
$$20x + 30y = 600$$

If you multiply each side of the first equation by 30, both equations will have 30y as a common term.

$$\text{Multiplication}$$

| $x + y = 25$ | $\Rightarrow$ | $30x + 30y = 750$ |
| $20x + 30y = 600$ | $\Rightarrow$ | $20x + 30y = 600$ |

Now you can use the subtraction method to eliminate 30y.

$$
\begin{array}{rcl}
30x + 30y = 750 & \Rightarrow & 30x + 30y = 750 \\
-(20x + 30y) = -650 & \Rightarrow & \underline{-20x - 30y = -650} \\
 & & 10x + 0 = 150
\end{array}
$$

The result is $10x = 150$, or $x = 15$. To find y, substitute 15 for x in the first equation. The result is $y = 10$.

Thus, Rachel needs to mix 15 gallons of the 20% solution and 10 gallons of the 30% solution to get 25 gallons of the 24% solution.

Critical Thinking To solve Rachel's system of equations by addition, what number should be used to multiply each side of the first equation?

EXAMPLE 2 Pick the Right Combination

Solve this system.

$$2x + 3y = 0$$

$$3x + 2y = -5$$

SOLUTION

One way to solve the system is to choose multipliers that will result in equal coefficients of x. Multiply the first equation by three and the second equation by two. Then eliminate a variable by subtraction.

$$
\begin{aligned}
3(2x + 3y) = 3(0) &\Rightarrow 6x + 9y = 0 &\Rightarrow & & 6x + 9y = 0 \\
2(3x + 2y) = 2(-5) &\Rightarrow 6x + 4y = -10 &\Rightarrow & & \underline{-(6x + 4y) = 10} \\
& & & & 0 + 5y = 10 \\
& & & & y = 2
\end{aligned}
$$

Substitute 2 for y and solve for x in the first equation.

$$
\begin{aligned}
2x + 3y &= 0 \\
2x + 3(2) &= 0 \\
2x &= -6 \\
x &= -3
\end{aligned}
$$

Check $(-3, 2)$ in the original equations.

Ongoing Assessment

Solve the system in Example 2 by first eliminating the variable y.

Critical Thinking The addition, subtraction, and multiplication methods are sometimes referred to as the elimination method. Explain why.

LESSON ASSESSMENT

1 How do you use addition or subtraction to solve a system of equations in standard form if neither of the variables are eliminated when you add or subtract?

2 Given a system of equations that is consistent and dependent, what happens when you try to solve the system using multiplication?

3 Given a system of equations that is inconsistent, what happens when you try to solve the system using multiplication?

4 Explain how division can be used to solve a system of equations.

Practice and Problem Solving

Solve each system by multiplication. Check the solution. If the system cannot be solved, identify the system as either inconsistent or dependent.

5. $3x - 4y = -9$
$x - 3y = 7$

6. $5x - 2y = 5$
$x + 4y = 12$

7. $2x + 3y = 4$
$4x + 5y = 6$

8. $2x + 3y = 4$
$8x + 2y = 5$

9. $2x - y = 6$
$3x - 2y = 2$

10. $4x + 3y = -2$
$3x + 2y = 3$

11. $4x + 3y + 6 = 0$
$3x - 2y - 4 = 0$

12. $8x + 5y = 3$
$7x + 3y = -7$

13. $6x + 9y = -3$
$2x + 3y = -1$

Write a system of equations to model each situation. Solve and check your solutions.

14. To service her car before a trip, Ms. Stein bought motor oil for $1.25 per quart and engine coolant for $2.50 a quart. Altogether she bought 10 quarts and spent $15. How many quarts of motor oil and coolant did Ms. Stein buy?

15. A rental company charged Mr. Juarez $90 to rent a generator and sprayer for six hours. On his next job Mr. Juarez rented the same equipment, but used the sprayer for 4 hours and the generator for 8 hours at a total cost of $100. What was the hourly rate for each piece of equipment?

16. Rasheed spent $15 for three rolls of ASA400 film and one roll of ASA100 film for his camera. He decided to take some extra film on a vacation. So Rasheed went back to the same store and bought two more rolls of ASA400 film and three more rolls of ASA100 film. The extra film cost $17. What is the cost of each type of film?

17. An airline baggage truck can hold 95 pieces of baggage in two separate compartments. In one trip from the airplane to the terminal, the truck is filled to capacity. One compartment is holding 14 more than twice the other. How many pieces of baggage are in each compartment of the truck?

18. The Giant Bike Company builds dirt bikes and racing bikes. The cost to build a dirt bike is $55 for materials and $60 for labor. The cost to build a racing bike is $70 for materials and $90 for labor. For this quarter, the Giant Bike Company has a budget of $9000 for materials and $10,500 for labor. How many of each type of bike can the company build this quarter?

Mixed Review

Write each number in scientific notation.
19. The wavelength of an X-ray is about 0.000000095 centimeters.

20. The mass of the Earth is about 6,600,000,000,000,000 metric tons.

21. A TV signal travels about 30,000,000,000 centimeters per second.

22. A protective coating used on a metal is about 0.0005 inches thick.

23. A formula used to determine the volume of a log is

$$V = \frac{1}{2} L(B + b)$$

The volume (*V*) of a log is 40 cubic feet. The length (*L*) of the log is 20 feet. The area (*B*) of one end of the log is 2.5 square feet. Find the area (*b*) of the other end of the log.

24. The force on a beam is found using this formula:

$$M = \frac{L(wL + 2P)}{8}$$

Rearrange the formula to solve for *P*.

Write an equation in one unknown to model each situation. Solve the equation and check your solution.

25. The total weight of two machine parts is 8.6 pounds. One part weighs 5 pounds more than the other part. Find the weight of each machine part.

26. Jack needs 23 two-by-four boards of 16-foot and 20-foot lengths to build rafters for a house. He needs seven fewer 16-foot boards than 20-foot boards. How many of each size does Jack need?

LESSON 8.6 SOLVING SYSTEMS WITH DETERMINANTS

In Lesson 8.5 you solved the system

$$2x + 3y = 0$$

$$3x + 2y = -5$$

using multiplication and subtraction. You can also solve this system by substitution or graphing. The last method you will learn to use is similar to the multiplication method.

First, write the coefficients as a square array of two rows and two columns enclosed in brackets. This array of numbers is called a **matrix**. The numbers or variables are called the **elements** of the matix. A matrix is usually designated by a capital letter. The following matrix is referred to as matrix *A*.

$$A = \begin{bmatrix} 2 & 3 \\ 3 & 2 \end{bmatrix}$$

The **determinant** of matrix A is computed in the following way:

$$(2 \cdot 2) - (3 \cdot 3) = -5$$

A pair of vertical lines is used to designate the determinant of a matrix. The determinant of matrix *A* is written as det *A*.

$$\det A = \begin{vmatrix} 2 & 3 \\ 3 & 2 \end{vmatrix} = -5$$

In general, if *a*, *b*, *c* and *d* are elements of a square matrix, the determinant of the matrix is computed as follows:

$$\begin{vmatrix} a & b \\ c & d \end{vmatrix} = ad - bc$$

The rule for finding the determinant of a square matrix with four elements is:

Find the difference of the products of the diagonal elements.

EXAMPLE 1 Finding a Determinant

Evaluate the determinant of matrix *A*.

$$A = \begin{bmatrix} 3 & 0 \\ 2 & -5 \end{bmatrix}$$

SOLUTION

Find the difference of the products of the diagonal elements of the determinant.

$$\det A = \begin{vmatrix} 3 & 0 \\ 2 & -5 \end{vmatrix}$$

The value of the determinant is $(3 \bullet -5) - (2 \bullet 0) = -15 - 0 = -15$.

Ongoing Assessment

Evaluate the determinant of matrix B.

$$B = \begin{bmatrix} 2 & 0 \\ 3 & -5 \end{bmatrix}$$

You can use determinants to solve for the values of x and y in a system of equations. First write the system with the equations in standard form.

$$ax + by = c$$
$$dx + ey = f$$

You can use the multiplication method to solve this system for x and y. The result is the following:

$$x = \frac{ce - bf}{ae - bd} \qquad y = \frac{af - cd}{ae - bd}$$

In the early 18th century a Swiss mathematician, Gabriel Cramer, devised a way to use determinants to solve a system of linear equations. To use Cramer's rule, you must first write the equations in standard form. The values for x and y can then be found by using determinants.

$$x = \frac{\begin{vmatrix} c & b \\ f & e \end{vmatrix}}{\begin{vmatrix} a & b \\ d & e \end{vmatrix}} \text{ and } y = \frac{\begin{vmatrix} a & c \\ d & f \end{vmatrix}}{\begin{vmatrix} a & b \\ d & e \end{vmatrix}}, \text{ where } \begin{vmatrix} a & b \\ d & e \end{vmatrix} \neq 0$$

Critical Thinking Evaluate the determinants in the numerators and denominators of the equations for x and y. Compare the results with the equations found above from the multiplication method. Are the equations the same?

EXAMPLE 2 Using Determinants

Solve the following system using determinants.

$$x - 4y = 1$$
$$2x + 3y = 13$$

Compare the equations to the general form:

$$ax + by = c$$
$$dx + ey = f$$

Find the corresponding values for a, b, c, d, e and f.

$$a = 1 \qquad b = -4 \qquad c = 1$$
$$d = 2 \qquad e = 3 \qquad f = 13$$

Substitute these values into the formulas for x and y, expressed as ratios of two determinants, as follows:

$$x = \frac{\begin{vmatrix} c & b \\ f & e \end{vmatrix}}{\begin{vmatrix} a & b \\ d & e \end{vmatrix}} = \frac{\begin{vmatrix} 1 & -4 \\ 13 & 3 \end{vmatrix}}{\begin{vmatrix} 1 & -4 \\ 2 & 3 \end{vmatrix}}$$

$$y = \frac{\begin{vmatrix} a & c \\ d & f \end{vmatrix}}{\begin{vmatrix} a & b \\ d & e \end{vmatrix}} = \frac{\begin{vmatrix} 1 & 1 \\ 2 & 13 \end{vmatrix}}{\begin{vmatrix} 1 & -4 \\ 2 & 3 \end{vmatrix}}$$

Finally, use the definition of a determinant to solve for x and y.

$$x = \frac{\begin{vmatrix} 1 & -4 \\ 13 & 3 \end{vmatrix}}{\begin{vmatrix} 1 & -4 \\ 2 & 3 \end{vmatrix}} = \frac{(1)(3) - (13)(-4)}{(1)(3) - (2)(-4)} = \frac{3 + 52}{3 + 8} = \frac{55}{11} = 5$$

Thus, $x = 5$.

$$y = \frac{\begin{vmatrix} 1 & 1 \\ 2 & 13 \end{vmatrix}}{\begin{vmatrix} 1 & -4 \\ 2 & 3 \end{vmatrix}} = \frac{(1)(13) - (2)(1)}{(1)(3) - (2)(-4)} = \frac{13 - 2}{3 + 8} = \frac{11}{11} = 1$$

Thus, $y = 1$.

The solution to the system of equations is the ordered pair $(5, 1)$.

Solve the following problem by setting up a system of linear equations and using determinants. A car dealer nets $500 for each family car sold and $1000 for each sports car sold. The dealer is urged to sell 4 family cars for each sports car sold. To stay in business, the dealer must net at least $3000 per week to meet costs. How many family cars and how many sports cars must the dealer sell each week?

ACTIVITY Using Technology

Most graphics calculators and some simple computer programs can evaluate determinants.

1 Explain how to find a determinant on the graphics calculator of your choice.

2 Find a computer program for solving a system using determinants. Program a computer to solve the system in Example 2.

WORKPLACE COMMUNICATION

The CELSAT Division has 7 technicians assigned. The Communications Electronics Division has 11. Write and solve a system of two linear equations to determine how many people each Division Chief can send to the course. What other factors might be considered in determining who attends the course?

At the weekly staff meeting, the Director says:

"I picked up this brochure at a training conference in Washington, DC last week. I want to send some of our top technical people from the CELSAT Division and the Communications Electronics Division to the course. The training budget will pay tuition and travel for nine people. The two Division Chiefs will decide who goes, but divide up the nine proportionately—depending on the number of technicians assigned to each Division."

The *Continuing Education Association* Announces a
One-Week Short Course on
Introduction to Modern Digital Communications

Time:	This course is offered Monday through Friday, the first full week of every month.
Place:	Headquarters, Continuing Education Association, Fairfax, VA
Instructors:	Ms. Joyce Morrison-Bell
	Mr. Raymond D. Wong
Course Outline:	The Electromagnetic Spectrum
	Principles of Communication
	Wireless Communication
	Satellite Communications
	Digital Data Communications
	Examples of Modern Communication Systems
	Hands-On Laboratory Activities
Who Will Benefit:	This course is designed for newly assigned technicians or experienced technical personnel needing a refresher course in modem applications of digital communications.
Prerequisites:	Successful completion of High School Algebra and Geometry. Knowledge of Trigonometry will prove useful.
Cost:	$1499
	Registration Form and Hotel Information Are Attached

CEA Announcement Number 32

LESSON ASSESSMENT

Think and Discuss

1 Explain how to use determinants to solve a system of equations.

2 What are the advantages and disadvantages of using determinants to solve a system of equations?

Practice and Problem Solving

Solve each system using determinants. If the system cannot be solved, identify the system as either dependent or inconsistent.

3. $y + 2x = -3$
$x - y = 1$

4. $3x + 5y = 2$
$2x - 3y = 5$

5. $2x - 3y = 12$
$5x + 2y = -8$

6. $7x + 6y = 5$
$5x + 4y = 4$

7. $2x - 5y = -5$
$15x - 6y = 15$

8. $3x + 8y = 0$
$6x + 7y = 5$

9. $2x - 4y = -14$
$3x = -7y + 5$

10. $3x + 4y = 4$
$9x + 12y = 16$

11. $6x + 3y = 1$
$3x - 2y = 4$

Write a system of equations to model each situation. Then solve the system using determinants and a calculator. Check the solution.

12. Home Security, Inc. monitors two separate radio signals from home security systems. The sum of the power levels of the signals is 4 milliwatts. The stronger signal has 10 times the power of the weaker signal. What is the power level of each signal?

13. A car travels 15 miles per hour faster than a truck. In the same amount of time that the car travels 195 miles, the truck travels 150 miles. Find the speed of each vehicle.

14. Lisa invests $10,000 in two different funds. One fund earns 5% annually, and the other earns 8% annually. At the end of one year, Lisa receives a check for $680 in earnings for her investment in the funds. How much did Lisa invest in each fund?

Mixed Review

Solve each equation for a.

15. $3a - 9 = -5a + 15$ **16.** $2(a + b) = c$ **17.** $\dfrac{a - 4}{9} = \dfrac{6}{15}$

Kim has recorded the number of video rentals for his store, Hello Hollywood, over a 14-day period.

125, 118, 126, 115, 221, 200, 101,
115, 135, 158, 140, 175, 218, 115

18. What is the mode of the data?

19. What is the median of the data?

20. Create a frequency distribution table where the first class is 101–125 videos rented and the remaining classes are in intervals of 25 videos rented.

The Smiths have four children, but you do not know the number of boys or girls.

21. Create a sample space to show the possible outcomes for boys and girls in the Smith family.

What is the probability that there are

22. Exactly three girls?   **23.** Exactly two boys?

24. At least two boys?   **25.** At least one boy and one girl?

MATH LAB

Equipment   Graphics calculator or graph paper

Problem Statement

You will solve linear equations by graphing the expressions on either side of the equal sign. This Activity assumes you are using a graphics calculator. Before you begin, be sure you know how to input an equation, determine the coordinates of a point, and use the ZOOM and TRACE features of your graphics calculator. If you do not have access to a graphics calculator, use graph paper to complete each step.

Procedure

a   Solve $3x - 7 = 11$ algebraically.

b   Graph $y = 3x - 7$ and $y = 11$ on the same coordinate axes. What are the coordinates of the point of intersection? How can you use these coordinates to find the solution to $3x - 7 = 11$?

c   Solve the equation $3x + 9 = 5x - 7$ algebraically.

d   Graph this system of equations.

$$y = 3x + 9$$
$$y = 5x - 7$$

What are the coordinates of the point of intersection? What is the solution to the system?

Consider this problem: Latoya has $30 and is saving additional money at the rate of $7 each week. Shawn has $50 and is saving at the rate of $3 each week. In how many weeks will each person have the same amount of money?

e   Set up a system of two equations to model the problem. Solve the system by graphing and check your results.

f You can solve the problem using one unknown. If x represents the number of weeks until each person has the same amount, you can model the problem with this equation:

$$30 + 7x = 50 + 3x$$

Solve this equation by graphing the expressions on each side of the equal sign.

g Compare the methods used in Steps **e** and **f**. How are they alike? How are they different?

h Solve the following problem using a graphics calculator. Then solve the problem algebraically using a system of equations. Which method do you prefer? Explain your answer.

The distance around a rectangular patio is 40 feet. The length is one and one-half times the width. Find the length and width of the patio.

Activity 2: Battleships and Mines

Equipment   String or cord, approximately 20 feet long
Paper (different color for each lab group)
Scissors
Masking tape
Tape measure
Optional: Graphics calculator

Problem Statement

You are navigating a battleship during wargames. Your course will take you across several enemy shipping lanes. Your mission is to lay mines at the points where your course crosses the enemy lanes. The enemy shipping lanes are represented by the following equations:

Enemy Lane 1: $x - y = -4$
Enemy Lane 2: $3x - y = 10$
Enemy Lane 3: $x - 2y = -2$

Procedure

a Clear an area of the classroom floor for use as a coordinate system. If your floor has square tiles, use the lines between the tiles as a grid for the *x*- and *y*-coordinate plane. If not, measure a uniform grid with a tape measure.

b As a class, identify a length that will serve as "one unit." On a tile floor, the length of one tile is very convenient. For floors without tiles, use a length of about 10 inches or 20 centimeters.

c This Lab will focus on only Quadrant I of the *xy*-coordinate system. Identify two perpendicular reference lines to serve as the positive *x*- and *y*-axes. For example, you can select two perpendicular rows of tiles near a corner of the classroom.

d Measure the units along each axis with the tape measure (or count tiles on a tile floor). Using small pieces of masking tape, label the units along your "axes" from $x = 0$ to $x = 20$ and $y = 0$ to $y = 20$. For convenience you may label every other unit. Create the coordinate system as a class. Only one coordinate system is needed for all the lab groups.

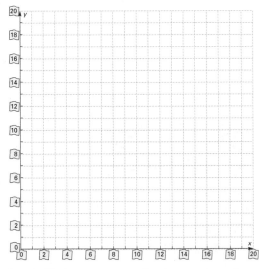

e Each group should use a different equation to represent the course of its battleship. Choose from the following equations.

Equation 1: $5x + 4y = 48$
Equation 2: $x + y = 18$
Equation 3: $3x + y = 10$
Equation 4: $15x - 200y = -1800$
Equation 5: $9x + 5y = 180$

f Determine the points where your equation (the path of your battleship) will intersect each of the enemy shipping lanes. Use any of the methods discussed in the text for solving a system of equations. Round your answers to the nearest 0.1 unit before they are recorded.

Points of Intersection	
Enemy Lane	*xy*-coordinates
1	
2	
3	

g Cut out three 1-inch squares of colored paper. Each square represents a "mine" dropped by your battleship. Each group should use a different color.

h For each point of intersection, locate the corresponding point on the classroom floor. Use masking tape to fasten a piece of colored paper to the floor at each point of intersection. This will indicate where your battleship has dropped a "mine."

i After all the lab groups have completed "dropping their mines," check each group's success. The equations for the shipping lanes are linear and represent lines. Find two ordered pairs on each of these lines. Use a string to connect the ordered pairs for each line. These three strings represent the three enemy shipping lanes. Did your "mines" lie within the shipping lanes?

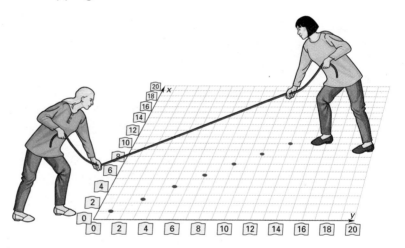

Activity 3: Unmixing a Mixture

Equipment Box of small bolts
Box of small nuts
Paper cup, approximately 8 ounces
Balance scale with one gram resolution

Problem Statement

Suppose you have a mixture containing two different items. You know the total count and the total weight of the mixture. You also know the weight of each type of item. But you do not know the number of each item in the mixture. You will determine how many items of each type are in the mixture with a simulation using nuts and bolts.

Procedure

a First determine and record the weight of one nut (N) and the weight of one bolt (B). If your scale is not accurate enough to measure just one nut, count and weigh 20 nuts. Divide this weight by 20 to find the weight of a single nut. Similarly, use 20 bolts to determine the weight of a single bolt.

b Place a "random" number of nuts and bolts into a cup. Do not attempt to put an equal number of each into the cup. However, you need to know the *total* number of objects in the cup.

c Give your cup to another lab group. Tell this group only the total number of objects in the cup.

d At the same time, you will receive a cup of nuts and bolts from another group. They should tell you the total number of items in the cup. Record this total as T. Do *not* remove the nuts and bolts from the cup.

e Use the scale to determine the total weight of the cup with nuts and bolts. Record this total weight W. Set the cup aside and perform the calculations that follow.

f Let x represent the number of nuts and y the number of bolts in the cup. Write an equation for the total number of nuts and bolts in the cup. Write an equation for the total weight of the nuts and bolts. (Can you safely ignore the weight of the paper cup?)

g From the preceding step, you should have two equations with two unknowns: x (the number of nuts) and y (the number of bolts). Solve these two equations for x and y. Check your solution by substituting back into your equations.

h Examine your solution from Step **g**. Did your mathematical solution result in whole numbers? If not, what whole number values for x and y would you choose as the actual numbers of nuts and bolts in the cup?

i Now count the nuts and bolts in the cup. Compare this to your solution for x and y. If the actual number of nuts and bolts is different than you predicted, explain why. Will the actual values for x and y make your equations true (or almost true)?

MATH APPLICATIONS

The applications that follow are like the ones you will encounter in many workplaces. Use the mathematics you have learned in this chapter to solve the problems.

Wherever possible, use your calculator to solve the problems that require numerical answers.

A motorboat takes 40 minutes to travel 20 miles down a river with the help of the current. On the return trip, the boat is powered at the same level and moves against the current. The upstream trip (the same 20-mile distance) now takes 50 minutes.

1 Let *x* represent the speed of the boat in the water and *y* represent the speed of the water current. While traveling downstream, the boat's speed—as seen from the shore—is the *sum* of the speed of the current and the speed of the boat in the water. Write an equation that describes the downstream speed. This sum is equal to the speed of 20 miles per 40 minutes, or 20/40 miles per minute.

2 While traveling upstream, the boat's speed—as seen from the shore—is the *difference* between the speed of the current and the speed of the boat in the water. Write an equation that describes the upstream speed. This difference is equal to the speed of 20 miles per 50 minutes, or 20/50 miles per minute.

3 Solve the system of two equations to find the speed of the boat in the water (*x*) and the speed of the current (*y*).

4 What are the units for the speeds you determined: Miles per hour? Miles per minute? Feet per minute? Explain.

You and your friends are planning a kayak trip. You are in charge of deciding which kayak rental company will be the least expensive. Company A charges $5 per hour for the use of its kayaks. Company B charges an $8 usage fee plus $3 per hour. If *c* represents the cost and *h* represents the hours used, you can write the following equations for each company:

Company A: $c = 5h$
Company B: $c = 3h + 8$

5 Draw a graph for each company showing cost as a function of hours.

6 Use the graph to find the number of hours where the costs of the two plans are equal.

7 If you and your friends are going to take a 6-hour trip, which plan is least expensive?

Bill and Laura mow lawns during the summer to earn money. They each calculate their start-up expenses, operating expenses, and income per hour of mowing. Bill and Laura write the following equations for their total income (y) after mowing for x hours:

$$\text{Bill:} \quad y = 9.10x - 125$$
$$\text{Laura:} \quad y = 9.10x - 100$$

8 Graph each equation on the same coordinate plane. Use x-values from zero to 20 hours.

9 Before Bill and Laura mow their first lawn, what is their total income? Explain your answer.

10 After mowing for 20 hours, what is each of their total incomes?

11 If both Bill and Laura work the same number of hours, when will Bill have the same total income as Laura? Explain your answer.

AGRICULTURE & AGRIBUSINESS

A farmer plants two kinds of crops on his 2500 acres of land. The income from Crop A is $230 per acre. The income from Crop B is $280 per acre. The farmer's goal is to earn $625,000 from the sale of the crops. How many acres of each crop should the farmer plant?

12 Let x be the number of acres of Crop A and y be the number of acres of Crop B. Equate sales of x acres of Crop A plus y acres of Crop B to the desired total sales of $625,000.

13 If the farmer uses all 2500 acres of available land, write an equation for the total number of acres planted.

14 Solve this system of equations to find the number of acres of each type of crop the farmer should plant to obtain the desired total sales.

15 Check your result by substituting the x- and y-values into your equations.

You have been offered a job marketing grain products. The company has two salary plans. Plan A pays a monthly salary of $700 plus a commission of $50 for each ton of grain products that you sell. Plan B pays a monthly salary of $500 plus a commission of $70 for each ton sold.

16 Write these two payment plans as a system of equations.

17 Draw a graph for each equation.

18 How many tons of grain must you sell for Plan A and Plan B to pay the same amount?

19 You estimate that you can sell about 25 tons per month. Which plan should you choose?

Roberta is selling two types of land. Type A land has access to water and sells for $900 per acre. Type B land does not have access to water and sells for $700 per acre. Roberta wants to earn $60,000 and will sell 75 acres.

20 Write an equation for the total earnings from the sale of land.

21 Write a second equation for the total number of acres Roberta will sell.

22 Solve this system of two equations by using determinants. How much of each type of land must Roberta sell to make $60,000?

23 Check your answers by substituting them into your original equations.

Mario is stocking a new pond for his fish farm. The pond is filled by a spring-fed creek at a rate that can support F fish:

$$F = 42m + 250$$

where m is the number of months the pond has been filling. Mario stocks the pond with an initial population of 110 fish (at month $m = 0$). He predicts a population growth rate of 8% per month. Using this information, Mario writes an equation for the population size P:

$$P = 110(1.08)^m$$

where P is the number of fish in the pond at month m, and m is the number of months since the fish were placed in the pond. Notice that this equation is *not linear*. This problem can be solved by graphing.

24 Graph both equations on a graphics calculator. Show values for m from zero months up to 50 months.

25 Mario predicts that the pond will take about 40 months to fill. He wants to begin harvesting the fish just before the population exceeds the limit that the pond can support. When will the predicted population of fish (*P*) equal the number of fish the pond can support (*F*)?

26 Should Mario wait till the pond fills to begin harvesting the fish, or should he start sooner?

BUSINESS & MARKETING

Chris is in charge of buying new suits for The Men's Shop. The purchasing budget is $20,000. Chris decides to stock the inventory with two different types of suits. One type is a designer label costing $400. The other is a nondesigner label costing $250. From previous sales records, Chris knows customers will buy nondesigner suits about 60% of the time and designer suits about 40% of the time.

27 Let *x* represent the number of nondesigner suits and *y* represent the number of designer suits. Equate the purchasing budget of $20,000 to the cost of purchasing *x* nondesigner suits and *y* designer suits.

28 To meet the sales demand, 60% of the purchases should be nondesigner suits and 40% should be designer suits. Write this as a proportion and solve the proportion for *y*.

29 Solve the system of equations from Exercises 27 and 28 by substitution.

30 How many of each type suit should Chris purchase to stay within her budget and meet the sales demand?

31 Check your result by substituting the *x*- and *y*-values into your equations.

You are starting a business that sells water filters. Your initial start-up cost is $1500, and each filter kit will cost $50. You plan to sell each unit for $75. From this information, you can write the following equations, where *n* is the number of filters sold:

$$\text{Cost} = 50n + 1500$$
$$\text{Income} = 75n$$

32 Write each equation as a function of *n*. Draw the graph of each function on the same coordinate plane.

33 From the graph, determine how many filters you need to sell to "break even." (That is, how many filters do you need to sell for income to equal cost?)

Susan's new company jet carries enough fuel for eight hours of flying when cruising at an airspeed of 220 miles per hour. Susan is heading west with a tailwind of 15 miles per hour. She needs to find the maximum distance west she can fly and still have enough fuel left to return home.

Westerly speed = 220 mph + 15 mph

Easterly speed = 220 mph – 15 mph

34 Consider the following variables: the time t_w for the westerly (outbound) flight and the time t_e for the easterly (return) flight. The total flight time is limited by the amount of fuel. Write an equation that shows the flight times, t_w and t_e, totaling eight hours.

35 Since the westerly flight is assisted by a tailwind, the airplane's speed is $220 + 15 = 235$ miles per hour. The distance traveled during the westerly flight is this speed times t_w. Write the product for the distance traveled during the westerly flight.

36 Since the easterly flight is into a headwind, the airplane's speed is $220 - 15 = 205$ miles per hour. The distance traveled during the easterly flight is this speed times t_e. Write the product for the distance traveled during the easterly flight.

37 Of course, Susan wants the outbound westerly flight distance to equal the easterly return flight distance. What happens if they are not equal? Equate the two products from Exercises 35 and 36 to obtain a second equation with t_w and t_e.

38 Solve the system of two equations. Check your results by substituting into your original equations.

39 Use the product from Exercise 35 to find how far west Susan can fly under these conditions and still have enough fuel to return home.

Your trucking company needs to move 21 tons of gravel. You have eight qualified drivers in the company and two types of trucks. One type of truck can haul 5 tons and the other type can haul 3 tons. An insurance requirement specifies that 5-ton trucks must have two drivers in the cab during operation. Three-ton trucks require only one driver. Let x be the number of 5-ton trucks you will use and y the number of 3-ton trucks.

40 Write two linear equations relating x and y that determine how many of each size truck is needed to move the gravel in one trip using all available drivers.

41 Solve the system using determinants.

42 How many of each size truck will you use?

You must determine the price schedule for concert seating at the local orchestra hall. You will sell tickets for two types of seats: one type will sell for $5 each, and the second type for $8 each. There are a total of 1500 seats, and you expect to sell tickets for all of them. Your ticket sales total $10,500. Let x be the number of seats that sell for $8 each and y be the number of seats that sell for $5 each.

43 Write an equation for the total number of seats as the sum of the two different types of seats.

44 Write an equation for the income from the sale of tickets ($10,500).

45 Solve this system of equations.

Amad wants to earn $500 on a $6000 investment. He is going to split his investment between stocks that he predicts will have an annual yield of 10% and a mutual fund he predicts will have a 7% annual yield. Let x be the amount invested in stocks and y be the amount invested in the mutual fund.

46 Write the equation for the total amount invested as the sum of the amount invested in stocks (x) and the amount invested in the mutual fund (y).

47 Write an equation that equates the amount that Amad wants to earn with the sum of the yield from the stocks (10% of the amount invested in stock) and the yield from the mutual fund (7% of the amount invested in the mutual fund).

48 Graph both equations on the same coordinate axes.

49 From the graph, find the amount that Amad should invest in stocks and the amount he should invest in the mutual fund. about

50 Solve the two equations using the method of determinants and compare the answer to the results obtained using the graphical method.

A cellular telephone company offers two service plans. The economy plan charges a $20 monthly fee, allows 30 minutes of free call time, and charges 50¢ per minute thereafter. The deluxe plan charges a $50 monthly fee, 25¢ per minute for call time, but no free minutes.

51 The cost of the economy plan (*CE*) can be written in terms of the number of minutes (*m*) of call time.

$$CE = 0.50\,(m - 30) + 20.00, \text{ for } m \geq 30$$

Multiply the terms in this equation. Rewrite it in slope-intercept form.

52 Write a similar equation for the cost of the deluxe plan (*CD*) in terms of the number of minutes (*m*) of call time.

53 Graph these two equations. Use the number of minutes of call time (*m*) as the horizontal axis and cost (*C*) as the vertical axis. What is the cost for the economy plan when *m* is less than 30 minutes?

54 Interpret the graph to find the number of minutes for which the two plans cost the same.

55 Explain how you can use the graph to decide which plan is better for you.

FAMILY AND CONSUMER SCIENCE

You are in charge of catering a banquet. To keep the costs down, you will serve only two entrees. One is a chicken dish that costs $5; the other is a beef dish that costs $7. There will be 250 people at the banquet, and the total cost of the food is $1500.

56 Let *x* be the number of chicken dishes you will prepare. Let *y* be the number of beef dishes. Equate the total number of entrees to the total number of people.

57 Write an equation that equates the total cost of the food to the cost for all the chicken dishes plus the cost for all the beef dishes.

58 Find the number of each type of entree you should prepare. Check your answers.

You are redecorating your home. You have decided to put down a combination of carpet and vinyl floor covering in the family room. The carpet costs $2 per square foot, and the vinyl covering costs $1 per square foot. You can spend a total of $500 on the materials. The area you want to cover is 300 square feet.

59 Let x be the number of square feet of carpet you will use. Let y be the number of square feet of vinyl. Equate the total cost of the materials to the cost of the carpet and the vinyl.

60 Write an equation that equates the total floor area (300 square feet) to the sum of the area covered by carpet and the area covered by vinyl.

61 Solve this system of equations to find the number of square feet of carpet and vinyl that meet your requirements. Check your answers.

You are shopping for refrigerators. Brand A costs $600 and uses $68 per year in electricity. Brand B costs $900 but uses only $55 per year in electricity.

62 Write a system of equations to model the total cost of each refrigerator over t years.

63 Assume that the domain of each equation is from zero years to 15 years (why not 70 years?). Draw a graph of the system.

64 Over the lifetime (that is, the domain) of these refrigerators, will the total costs ever be the same? If so, when?

65 Which refrigerator has a lower total cost over a 15-year lifetime?

As an interior decorator, you are selecting the lighting for a client's new family room. You must decide between regular incandescent light bulbs and energy-saving fluorescent light bulbs. The incandescent bulbs cost 50¢ each and will last for 1000 hours. The fluorescent light bulbs cost $11 each and last 9000 hours. However, the incandescent bulb costs $0.006 per hour to operate, while the fluorescent bulb operates for only $0.0018 per hour. You want to show your client that you have chosen the more economical lighting system.

66 Write a system of equations for the total cost of each type of light bulb for equal time periods. (Hint: one fluorescent bulb lasts as long as nine incandescent bulbs.)

67 Graph the two equations for the total cost for each bulb over the span of 9000 hours.

68 Is there a solution to these two equations? That is, will the total cost of the incandescent bulbs ever equal the total cost of the fluorescent bulb? If so, when?

69 Based on cost alone, which bulb would you recommend to your client? Explain your reasoning.

HEALTH OCCUPATIONS

Shon needs three liters of an 8% saline solution. He has a 5% saline solution and a 9% solution in the lab stock room. Before he mixes the two solutions, he needs to calculate the right proportions. Let x be the number of liters of the 5% solution and y be the number of liters of 9% solution.

70 Equate the sum of x and y to the total number of liters needed.

71 To obtain a second equation, Shon starts with the word equation below. The equation uses the fact that the amount of *salt* in the saline solution is the concentration of the solution times the number of liters of solution. When Shon mixes solution A and solution B, the amount of salt in the mixture is the amount of salt from solution A plus the amount of salt from solution B.

(Concentration of A)(Liters of A) + (Concentration of B)(Liters of B) = (Concentration of mixture)(Liters of mixture)

Substitute the appropriate variables and numbers into the word equation and write Shon's second linear equation.

72 Use the method of substitution to solve the system of two equations. How many liters of 5% and 9% saline solution should Shon mix to obtain 3 liters of 8% saline solution? Check your answers.

INDUSTRIAL TECHNOLOGY

Fahrenheit and Celsius are two different scales of temperature measurement. Linear equations relate the two scales.

$$F = aC + b$$

F = the temperature in degrees Fahrenheit.
C = the temperature in degrees Celsius.
a and b = the constants to be determined.

The two scales are related by the fact that water freezes at 32°F or 0°C. Similarly, water boils at 212°F or 100°C.

73 Substitute the temperatures for freezing water into the equation above to obtain an equation with unknowns a and b.

74 Substitute the temperatures for boiling water into the equation above to obtain a second equation with unknowns a and b.

75 Solve this system of equations to determine the values of a and b that satisfy both equations.

76 Rewrite the equation $F = aC + b$ with the calculated values of a and b. Check the resulting equation for $C = 0$ and $C = 100$.

Laura needs 4000 pounds of a 17% copper alloy. To make the alloy, she will mix a 23% copper alloy and a 12% copper alloy. Let x be the number of pounds of the 23% copper alloy. Let y be the number of pounds of the 12% copper alloy.

77 Write the equation for the total number of pounds needed for the mixture as the sum of the pounds of 12% alloy and the pounds of 23% alloy.

78 To obtain a second equation, Laura starts with the word equation below. The equation uses the fact that the amount of copper in an alloy is the concentration of the alloy times the number of pounds of alloy. When Laura mixes the two alloys, the amount of copper in the mixture is the amount of copper from one alloy plus the amount of copper from the other alloy.

(Concentration of A)(Quantity of A) + (Concentration of B)(Quantity of B) = (Concentration of mixture)(Quantity of mixture)

Rewrite this word equation using the variables x and y. Remember to express the percentages as decimal values.

79 Use the method of substitution to solve the two equations. How many pounds of 23% and 12% alloy should Laura mix to obtain 4000 pounds of 17% alloy? Check your answers by substituting the results for x and y into your equations.

You are calibrating a new thermocouple that is used for measuring temperatures between 0°C and 100°C. To do this, you must be able to relate the thermocouple voltage to the temperature being measured. You know that the thermocouple's output voltage is roughly linear between 0°C and 100°C (that is, you can write a calibration equation in the form $y = mx + b$). Thus, for a temperature T, the thermocouple should produce a voltage V according to this calibration equation.

$$T = Vm + b$$

Two of the test measurements with your new thermocouple show that when $T = 0$°C, $V = -1.56$ millivolts, and when $T = 100$°C, $V = 4.76$ millivolts.

80 Substitute each pair of temperature and voltage measurements into the calibration equation to obtain a pair of equations with unknowns m and b.

81 Solve for m and b. Check the results by substituting your answers into the two equations.

82 Write the correct calibration equation with the values of m and b you determined.

Kirchhoff's loop rule analyzes the current (I) in an electronic circuit. This rule states that in any closed loop of a circuit, the sum of the voltage drops (across the resistors) must equal the sum of the voltages supplied. If you apply this rule to the circuit shown here, you obtain the following equations:

$$12 - 12I_1 - 7(I_1 + I_2) = 0$$

$$16 - 22I_2 - 7(I_1 + I_2) = 0$$

83 Simplify each equation to obtain a system of equations in standard form.

84 Solve this system of equations for I_1 and I_2. Check your solution.

Skills

Solve each system by substitution.

1. $x = -3y + 8$
$2x - 5y = -17$

2. $x - 2y = -4$
$5x - 3y = 1$

3. The sum of two integers is 33. One of the integers is 3 more than the other. What are the integers?

Solve each system by multiplication.

4. $2x - y = -7$
$3x + y = -3$

5. $2x - y = 1$
$4x - 3y = -5$

Applications

6. The perimeter of a rectangle is 48 feet. The width is 6 feet less than the length. Find the dimensions of the rectangle.

Use the following system of equations for Exercises 7-9.

$$x + y = 3$$
$$2x - 3y = 16$$

7. Write the ratio of determinants needed to solve for x.

8. Write the ratio of determinants needed to solve for y.

9. Solve the system using determinants.

Write a system of equations to model each situation. Solve the system by any method you choose.

10. The total cost for printing a book includes a setup fee and a charge for each book printed. The total cost to print 500 books is $1750. The total cost for printing 1000 books is $3250. What is the setup fee?

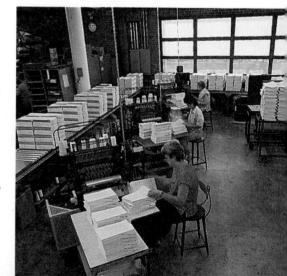

11. A car dealer makes a profit of $750 for each 4-door family car sold and $1250 for each 2-door sports car sold. The dealer needs to sell three times as many family cars as sports cars each week. How many of each car must the dealer sell to make a profit of $7000 each week?

12. A chemist needs a 5% sodium solution. The chemist has a 4% sodium solution and an 8% sodium solution available. How much of each solution should be mixed to make 400 milliliters of 5% sodium solution?

Math Lab

13. When solving a system of two equations graphically, what point represents the solution to the system?

14. The course of your battleship follows $3x + y = 10$. The enemy's shipping lanes follow two equations:

Enemy Lane 1: $3x - y = 10$
Enemy Lane 2: $6x + 2y = 10$

Can you lay mines to intercept both enemy lanes? Why or why not?

15. The weight of a bolt is measured at 125 gm. The weight of a nut is measured at 35 gm. A paper cup that contains a total of 29 nuts and bolts is measured at 2190 gm. How many nuts and how many bolts are in the cup?

CHAPTER 9

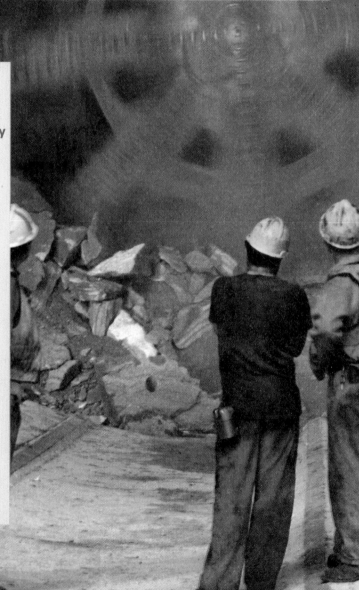

People in today's high tech world demand quality products. To meet these expectations, engineers design products to high standards using very tight tolerances. These tolerances are often expressed as inequalities. Learn the concepts in this chapter and you will be another step closer to a bright and profitable future.

INEQUALITIES

OBJECTIVES

1. Solve linear inequalities in one variable and graph their solutions.
2. Solve combined inequalities.
3. Solve inequalities involving absolute values and graph their solutions.
4. Graph linear inequalities in two variables.
5. Solve linear programming problems.

So far you have learned to translate work-related problems into equations and then solve the equations for unknowns. However, there are many workplace problems that involve unequal expressions or inequalities. For instance, in building the "Chunnel," technicians used precise tolerances to ensure the tunnel stayed on course as it was constructed under the English Channel. Engineers stated these tolerances to the technicians using such phrases as "greater than" or "less than." These phrases are also used to express inequalities.

For instance, suppose you need to know if a $\frac{3}{4}$-inch bolt is longer than a $\frac{5}{8}$-inch bolt. A number line can help you choose the longer bolt. Visualize a number line divided into eight parts.

You can quickly determine that $\frac{3}{4}$ is farther to the right on the number line than $\frac{5}{8}$. Thus, $\frac{3}{4}$ is greater than $\frac{5}{8}$, and the $\frac{3}{4}$-inch bolt is longer. In this chapter, you will use number lines to picture inequalities and explain their behavior.

As you watch the video for this chapter, notice how people use inequalities in the workplace.

LESSON 9.1 INEQUALITIES AND THE NUMBER LINE

The statement "five is greater than three" is an example of an **inequality**. You can write this statement using a *greater than* symbol.

$$5 > 3$$

You will use several symbols to describe inequalities.

greater than	$>$
greater than or equal to	$\geq$
less than	$<$
less than or equal to	$\leq$
not equal to	$\neq$

An inequality statement can be true, false, or conditional. The inequality $5 < 3$ is false. The inequality $-5 < 3$ is true. However, you cannot tell if the inequality $x < 3$ is true until you know what replacements to use for x. The inequality $x < 3$ is a *conditional inequality*.

You can specify an **interval** by combining conditional inequalities. For example, let n represent a member of the set of whole numbers in the interval between 5 and 10. This set is $\{6,7,8,9\}$. The set can be specified with the **compound inequality**

$$5 < n < 10$$

This inequality is read

$$5 < n \quad \text{and} \quad n < 10$$

and means that n is *between 5 and 10.*

The whole numbers (6, 7, 8, and 9) that make this inequality true are the **solutions** of the inequality. If the domain is the set of all real numbers, then there are an infinite number of solutions because there are an infinite number of real numbers between 5 and 10. In this case, you can use a graph to show the solution.

The graph of the inequality $5 < x < 10$ is called an **open interval**. An open interval does not include the endpoints. Notice that the endpoints of an open interval are represented by an open circle. If the solution *does* contain either endpoint, the corresponding circle is solid. If the solution contains both endpoints, the graph is called a **closed interval**.

Critical Thinking Describe the graph of $2 \le x \le 3$.

EXAMPLE 1 Graphing Inequalities

Graph the solution for each inequality.

a. $x < 35$ **b.** $x \le -10$ **c.** $x > 25$ **d.** $0 \le x \le 20$

SOLUTION

a. $x < 35$

b. $x \le -10$

c. $x > 25$

d. $0 \le x \le 20$

Ongoing Assessment

Graph the solution to each inequality.

a. $x > -3$ **b.** $x < 5$ **c.** $-2 \le x \le 2$

EXAMPLE 2 Writing an Inequality

Write an inequality to describe each graph.

a.

b.

c.

SOLUTION

a. $x > 2$ **b.** $-3.5 \le x \le 1.5$ **c.** $x \ge -5$

Many businesses' financial situations are modeled by inequalities.

EXAMPLE 3 Graphing an Interval

Every year since 1996, the Gorgeous Garment Company has made a profit of at least $20,000. It has never made a profit greater than $50,000. Graph an inequality to represent the Gorgeous Garment Company's profit since 1996.

SOLUTION

Let x represent the profit since 1996. The Gorgeous Garment Company's profit is described by the inequality

$$\$20{,}000 \leq x \leq \$50{,}000$$

The graph is as follows:

```
←——+——+——+——+——+——+——+——+——+——+■——■——+——+——+——→
  -80 -70 -60 -50 -40 -30 -20 -10  0  10 20 30 40 50 60 70 80
```
Scale in thousands of dollars

CULTURAL CONNECTION

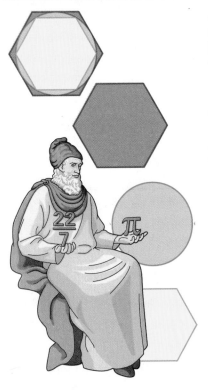

Archimedes, one of the greatest of all Greek mathematicians, lived in Egypt more than 2000 years ago. Using only the Egyptian numerals available in his time, he performed complex computations with whole numbers and fractions. Archimedes was probably the first person to compute π correctly to two decimal places. He started with two regular hexagons, one drawn inside and one drawn outside a circle, and the knowledge that π is the ratio of the circumference of a circle to its diameter.

Archimedes computed the perimeters of the two hexagons. He then formed two ratios. The first ratio was the perimeter of the outer hexagon to the diameter of the circle.

The second ratio was the perimeter of the inner hexagon to the diameter of the circle. Archimedes knew the value of π must be between the values of these two ratios. Then he began doubling the number of the sides of the inside and outside polygons to get values closer and closer to π. He finally figured out that π is greater than $\frac{223}{71}$ and less than $\frac{22}{7}$.

1. Write an inequality to model Archimedes' approximation for π.

2. Change the fractions to decimals.

3. How do Archimedes' ratios compare to the value of π we use today?

LESSON ASSESSMENT

Think and Discuss

Let *a* and *b* represent two real numbers. Explain how to graph each inequality.

1 $x < a$

2 $x > b$

3 $a \le x \le b$

Practice and Problem Solving

Write an inequality to describe each graph.

4.

5.

6.

7.

8.

9.

Graph each inequality.

10. $x < 5$ **11.** $x > 2$

12. $x < -4$ **13.** $x \ge -6$

14. $-2 \le x \le 2$ **15.** $4 < x < 7$

Write an inequality that models each situation.

16. Paul's monthly income is more than $1500.

17. Angie must drive less than 25.5 kilometers.

18. The winning time in the 100-meter dash was at least 10 seconds.

19. It takes between $1\frac{1}{2}$ and 2 hours to play a soccer game.

Mixed Review

An antique phonograph needle is approximately a smooth half sphere. Suppose one needle is 0.02 centimeters in diameter. A measurement of the tracking force on a phonograph yields 0.156 Newtons. Write your answers to Exercises 20–21 in scientific notation.

20. The surface area of a sphere is $4\pi r^2$. Convert the diameter of the phonograph needle half sphere to meters. What is the surface area of the needle tip in square meters?

21. Divide the tracking force by the area of the needle tip to find the pressure exerted by the needle on the phonograph record in Newtons per square meter.

The Air Care Furnace Company charges $50 for each service call plus $25 per hour for labor.

22. Write a function to model the cost of having Air Care repair a furnace at someone's home.

23. Graph the function.

24. What are the slope and y-intercept of the graph?

LESSON 9.2 THE ADDITION PROPERTY OF INEQUALITY

The cost of a matinee ticket is $3. After you pay for the ticket, you have less than $9 left. How much did you have before you bought the ticket?

Model this situation with an inequality. Let x represent your initial amount of money.

$$x - 3 < 9$$

How can you solve the inequality?

ACTIVITY Exploring Inequalities

1 Draw a number line from -8 to 8.

2 Circle the point -4. Circle the point 2. An inequality describing the relationship between the coordinates you circled is $-4 < 2$.

3 Moving to the right on the number line represents addition. Move three points to the right of each point. Circle the new coordinates.

4 Add 3 to each side of the inequality in Step 2. What is the new inequality showing the addition of 3 to each side of the original inequality? Find the sum of each side of the inequality.

5 Moving to the left on the number line represents subtraction. Start again with the inequality in Step 2. This time move three points to the left. Circle the new coordinates.

6 Subtract 3 from each side of the inequality in Step 2. What is the new inequality showing the subtraction of 3 from each side of the original inequality? Find the difference on each side of the inequality.

7 Examine the new coordinates from Steps 3 and 5. Is the inequality still true when you add or subtract the same number from each side?

In the Activity, you modeled two properties of inequalities. These properties are the same as the properties you have used to solve equations.

Addition Property of Inequality
If the same number is added to each side of an inequality, the resulting inequality is still true.

Subtraction Property of Inequality
If the same number is subtracted from each side of an inequality, the resulting inequality is still true.

Now you can use the Addition Property of Inequality to solve the matinee ticket problem.

$$x - 3 < 9 \qquad \text{Given}$$

$$x - 3 + 3 < 9 + 3 \qquad \text{Addition Property of Inequality}$$

$$x < 12 \qquad \text{Simplify}$$

Thus, you started with less than $12.

EXAMPLE 1 Using the Subtraction Property of Inequality

Solve $a + 4 \geq -1$. Graph the solution.

SOLUTION

Use the Subtraction Property of Inequality to subtract 4 from each side of the inequality.

$$a + 4 \geq -1 \qquad \text{Given}$$

$$a + 4 - 4 \geq -1 - 4 \qquad \text{Subtraction Property of Inequality}$$

$$a \geq -5 \qquad \text{Simplify}$$

Solve and graph each inequality.

a. $8 + m \le 5$ **b.** $15 > s - 9$

EXAMPLE 2 Solving an Inequality

Juanita has at least $270 in her savings account after she makes a deposit of $25. How much did Juanita have in her savings before the deposit?

SOLUTION

Words like "at least" and "at most" lead to statements of inequality. Since Juanita has at least $270, she can have $270 or more in her account. Her savings is *greater than or equal to* $270. Model the problem with an inequality and solve. Let x represent the amount of money before the deposit.

$x + 25 \ge 270$	Given
$x + 25 - 25 \ge 270 - 25$	Subtraction Property of Inequality
$x \ge 245$	Simplify

Juanita had at least $245 in her savings before she made the deposit.

Critical Thinking   Suppose a, b, and c are real numbers. Use these variables to write the Addition Property of Inequality and the Subtraction Property of Inequality as algebraic statements.

LESSON ASSESSMENT

Think and Discuss

1 How is the Addition Property of Inequality used to solve inequalities such as $x - a < b$? Give an example.

2 How is the Subtraction Property of Inequality used to solve inequalities such as $x + a < b$? Give an example.

3 Replace the ? in x __?__ 10 with an inequality sign to represent

a. at most   **b.** more than
c. less than   **d.** at least   **e.** no more than

Solve each inequality. Graph the solution.

4. $x + 5 < 7$ **5.** $x + 2 > -4$ **6.** $x - 3 \geq 4$

7. $x + 4 > -1$ **8.** $x + 3 < -5$ **9.** $x - 5 < 1$

10. $x + 2 \leq 2$ **11.** $x - 3 \geq -2$ **12.** $x - 1 \leq 4$

Write and solve an inequality that models each situation.

13. Tai has saved $35 toward the purchase of a stereo. She needs to save at least $150 by the end of the month to buy the stereo on sale. How much more does Tai need to save to purchase the stereo at the sale price?

14. You can take at most 100 pounds of luggage on an airplane. One of your bags weighs 55 pounds. How many more pounds of luggage can you take?

15. The difference between two numbers is less than 50. The smaller number is 15. What values can the larger number be?

16. State University accepts no more than 4500 students into its freshman class. So far 2000 students have been accepted. How many more students can State University accept?

17. The Home Town Gift Shop plans to spend no more than $350 on advertising in May. The shop has already spent $210. How much more can the shop spend on advertising?

18. At least 30,000 people attended a rally. This is 7500 more than attended the rally last year. How many people attended the rally last year?

Mixed Review

The daily production rates at an electronics assembly plant were recorded for 16 days:

42, 44, 38, 38, 43, 37, 44, 41
38, 46, 41, 37, 44, 42, 44, 37

19. Find the mode of the data.

20. Find the median of the data.

21. Find the mean of the data.

22. The president of the plant wants to know the average daily production rate. Which measure of central tendency would you use to answer this question?

A car lot has 20 sedans, 2 convertibles, 13 pickup trucks, and 11 sports cars for sale. Suppose each vehicle has an equally likely chance of being sold.

23. What is the probability that the first car sold is a convertible?

24. What is the probability that the first car sold is a sedan or a sports car?

25. Two cars were sold at once. What is the probability that both cars were convertibles?

26. A customer bought one of the vehicles. He brought it back and then bought one of the other vehicles. What is the probability that he bought a pickup first and a sports car second?

LESSON 9.3 THE MULTIPLICATION PROPERTY OF INEQUALITY

In Chapter 4, you used two methods to solve multiplication equations such as $2x = 12$. You found the solution by dividing each side by 2 or by multiplying each side by $\frac{1}{2}$. Either method results in the same solution, 6.

Can you use the same methods to solve the inequality $2x < 12$?

ACTIVITY 1 **Multiplying and Dividing by a Positive Number**

1 Draw a number line from −8 to 8.

2 Circle the points −4 and 2. Write an inequality describing the relationship between the coordinates you graphed.

3 Multiply each coordinate by 2. Plot the new coordinates on the number line.

4 Multiply each side of the inequality in Step 2 by 2. What is the new inequality showing the multiplication of 2 to each side of the original inequality. Now find the products on each side of the inequality.

5 Examine the new coordinates. Is the inequality still true when you multiply each side by the same positive number?

6 Repeat Steps 3–5. This time divide by 2.

7 What is the solution of $2x < 12$?

So far, it appears that solving a multiplication *inequality* is just like solving a multiplication *equality*. However, there is one case that is different. What happens when you solve $-2x < 12$?

ACTIVITY 2 **Multiplying and Dividing by a Negative Number**

1 Draw a number line from −8 to 8.

2 Circle the points −4 and 2. Write an inequality describing the coordinates you circled.

3 Multiply each coordinate by −2. Plot the new coordinates on the number line.

4 Rewrite the inequality in Step 2 to show each coordinate multiplied by −2. Find the products on each side of the inequality. Be careful to use the correct inequality sign.

5 Examine the new coordinates. What must you do to make the inequality true when you multiply each side of an inequality by the same negative number?

6 Repeat Steps 3–5. This time divide by −2.

7 What is the solution of $-2x < 12$?

The following chart summarizes the findings you made in the two Activities:

Given Inequality	Algebraic Operation Involved	Resulting Inequality	Observation
$-4 < 2$	multiply by 2	$-8 < 4$	Inequality is true with *same* symbol
$-4 < 2$	divide by 2	$-2 < 1$	Inequality is true with *same* symbol
$-4 < 2$	multiply by −2	$8 > -1$ (note: symbol has reversed)	Inequality is true only if < *changes to* >
$-4 < 2$	divide by −2	$2 > -1$ (note: symbol has reversed)	Inequality is true only if < *changes to* >

The inequality sign shows order. When you multiply or divide each side of an inequality by a *positive* number, the order remains the

> > > ➡ × or ÷ by a negative number ➡ < < <

same. When you multiply or divide each side of an inequality by a *negative* number, the order is reversed. These statements about inequality are combined as follows:

Multiplication and Division Properties of Inequality

For all numbers a, b, and c:

If $a < b$ and $c > 0$, then $ac < bc$ and $\frac{a}{c} < \frac{b}{c}$.

The order remains the same.

If $a < b$ and $c < 0$, then $ac > bc$ and $\frac{a}{c} > \frac{b}{c}$.

The order is reversed.

EXAMPLE 1 Solving an Inequality

Solve and graph $-3x + 1 \geq -2$.

SOLUTION

$-3x + 1 \geq -2$	Given
$-3x + 1 - 1 \geq -2 - 1$	Subtraction Property of Inequality
$-3x \geq -3$	Simplify
$\dfrac{-3x}{-3} \leq \dfrac{-3}{-3}$	Division Property of Inequality (reverse the inequality sign)
$x \leq 1$	Simplify

Ongoing Assessment

Solve and graph each inequality.

a. $5n + 3 < 3n + 8$ **b.** $2(3 - 2n) \geq 12$

EXAMPLE 2 The Cost of Doing Business

To fix a pool heater, a pool service company estimates a charge of at least $138. This price includes a service charge of $40 and an hourly rate of $28. Write and solve an inequality to determine how many hours the pool service estimates the repair will take.

Let x represent the number of hours. Write the inequality to model the problem.

$$40 + 28x \geq 138$$

Solve the inequality.

$40 + 28x \geq 138$	Given
$28x \geq 98$	Subtraction Property
$x \geq 3.5$	Divide each side by 28

The pool service estimates the job will take at least three-and-a-half hours.

WORKPLACE COMMUNICATION

What exercise rates for the stationary bike and stair-climber do you think you could sustain? At these rates, how much time on each machine would meet or exceed the recommended amount of exercise? Some people prefer to use the bike for half the time of a workout and the stair climber for the other half. Under this condition, and at the same exercise rates as before, how much time on each machine would meet or exceed the recommended amount of exercise?

mw Modern Workplace, Inc.

Memo to: All MWI Employees
From: MWI Medical Clinic

I am pleased to announce the availability of new exercise machines in MWI's health and fitness center. Six new stationary bikes and six stair-climbers are now available for your aerobic exercise program. The fitness center is open from 5 AM to 8 PM to help you burn calories before work, during lunch, or after hours.

The new machines are popular forms of aerobic exercise, and they are effective. The number of calories you burn using the machines, or any other form of exercise, depends on a number of factors, including your physical size and fitness level. But you can estimate the calories burned on the stationary bike and stair-climber with the chart below, which was provided by the machines' manufacturer. As a guideline for your exercise program, I recommend a sustained activity level that burns at least 300 calories. I also recommend you exercise at least three times per week.

Numerous studies have shown that regular exercise results in a healthier, happier, less stressful, and more productive workplace. If you are not regularly exercising now and would like to begin, please start with a visit to my office. I will give you information on exercise and diet, and, if needed, I can do a simple cardiovascular screening.

If you need an appointment, please call Vickie at extension 6-7893. I look forward to seeing you at the fitness center.

S.C. McMahan

Sandra C. McMahan
Physician's Assis
Head, MWI Med

Stationary Bike		Stair-Climber	
Revolutions per Minute	Calories per Minute	Steps per Minute	Calories Burned per Minute
70	6	70	8
80	8	80	10
90	12	90	15
100	16	100	20
110	18	110	23

LESSON ASSESSMENT

Think and Discuss

1 What happens to the order of the inequality when you multiply or divide each side of the inequality by a negative number?

2 What are the similarities and differences between solving inequalities and equalities?

Practice and Problem Solving

Solve each inequality. Graph the solution.

3. $3x < 6$

4. $4x < -8$

5. $\dfrac{x}{2} \geq -5$

6. $-2x < 7$

7. $-\dfrac{3}{4}x \geq 12$

8. $3x - 12 < 0$

9. $4x + 3 \geq -5$

10. $3x < -x - 5$

11. $5x - 5 > -9 + 3x$

Write and solve an inequality for each situation.

12. The perimeter of a rectangular patio is at most 68 feet. The length is 2 feet more than the width. What are the largest possible dimensions of the patio?

13. Margarita is choosing from among several jackets on sale for 30% off. All the jackets are originally over $75. What is the least amount Margarita can save on one jacket?

14. Lee's parents have agreed to give him an interest-free loan to buy a used car. Lee wants to borrow less than $5000. He plans to make 24 equal payments. What is the largest payment he will make?

15. Mr. Lewis is staying in a hotel for eight nights during a sales conference. He can spend no more than $1000 for lodging. What is the most Mr. Lewis can pay for each night's lodging?

16. Sarah charges $15 per hour to repair electronics equipment. Her goal is to earn at least $540 per week. So far this week, she has earned $450. What is the least number of hours Sarah must work during the remainder of this week to meet her goal?

17. Kelli needs at least 270 total points on three tests to earn an A. On her first two tests, she scored 86 and 88. What is the lowest score Kelli can make and still earn an A?

18. Wayne wrestles in the 148 pound division. He is training for a meet that is five days away. If Wayne's current weight is 154.5 pounds, what must be his least average weight loss (in pounds per day) for him to wrestle in his division?

19. John has $20 to buy gas for his truck. He needs to have at least $5 remaining after he pays the bill. If gas costs $1.39 per gallon, what is the greatest number of gallons of gas John should buy?

Mixed Review

A manufacturing company uses the linear function $y = 3x + 5$ to model the cost (y) of producing x gears.

20. Name the independent variable.

21. Find the slope of the graph.

22. Find the y-intercept of the graph.

23. Find the function that results when x is replaced by $x + 2$.

The same company uses the nonlinear function $y = \sqrt{x + 3} - 2$ as coordinate points in the computer-aided design of one of the gears.

24. What are the domain and range of the function?

25. What is the value of y when x is replaced by 1?

26. Graph the function.

LESSON 9.4 SOLVING COMPOUND INEQUALITIES

Inequalities are often combined into one statement. Consider these two inequalities:

$$2 < 3 \quad \text{and} \quad 4 < 5 \qquad 7 > 4 \quad \text{or} \quad 6 > 1$$

The first inequality states that "2 is less than 3 *and* 4 is less than 5." The second states that "7 is greater than 4 *or* 6 is greater than 1." Such inequalities are called **compound inequalities.** Compound inequalities are often used in the workplace.

Todd is a testing supervisor in a factory that produces and packages small metal parts. A packaging machine is set to drop 500 nails into each box. However, the nails vary slightly in weight. A 5-pound box of nails is approved for shipping if its weight is *between* 78.5 ounces and 81.5 ounces. To model the tolerance, Todd uses the inequality

$78.5 < x < 81.5$, where x is the weight of the nails and box

The inequality $78.5 < x < 81.5$ is an example of a compound inequality. It combines the two simple inequalities

$$78.5 < x$$
$$x < 81.5$$

connected by the word *and*. The inequality is read:

78.5 is less than *x and x* is less than 81.5

Conjunctions

A compound inequality that uses the word *and* to separate two simple inequalities is an example of a **conjunction**. The solution set of a conjunction is the set of solutions that make *both* simple inequalities true. This solution is the intersection of the solutions of the simple inequalities.

EXAMPLE 1 Solving a Conjunction

Solve and graph $-3 < (2x - 3) \le 5$.

Write the simple linear inequalities separately. Solve each inequality.

$$-3 < (2x - 3) \qquad and \qquad (2x - 3) \leq 5$$
$$-3 + 3 < (2x - 3) + 3 \qquad and \qquad (2x - 3) + 3 \leq 5 + 3$$
$$0 < 2x \qquad and \qquad 2x \leq 8$$
$$0 < x \qquad and \qquad x \leq 4$$

The solution is the set of all numbers greater than zero and less than or equal to four, which is

$$0 < x \leq 4$$

To graph the conjunction, graph the common solution to the two simple inequalities on the same number line.

The solution set for the conjunction is the intersection of the solution sets of the two simple inequalities. This is shown as the shaded segment that is common to both graphs.

There is a simple method to check your solution.

1. Choose a value in the intersection and substitute it into each simple inequality. The value should make both of the original inequalities true.

2. Choose a value to the left of the interval. This value should make one of the simple inequalities true.

3. Choose a value to the right of the interval. This value should make the other simple inequality true.

Critical Thinking How can you check the solution set of a conjunction by using the endpoint values of each simple inequality?

Ongoing Assessment

Solve and graph each conjunction. Check your solution set.

a. m is greater than -5 and less than 4 **b.** $-2 \leq 3y + 1 \leq 5$

The solution set of a conjunction is the **intersection** of two sets of numbers. The intersection of two sets consists of the numbers that are common to both sets; in other words, the numbers in one set that are also in the other set. The shaded area in the Venn diagram below represents the intersection of sets A and B. The intersection of the two sets is designated A ∩ B.

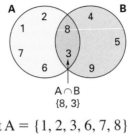

A∩B
{8, 3}

Set A = {1, 2, 3, 6, 7, 8}
Set B = {3, 4, 5, 8, 9}

Since only the elements 3 and 8 are in the shaded (overlap) region,

$$A \cap B = \{3, 8\}$$

Consider again the combined inequality in Example 1:

$$-3 < (2x - 3) \leq 5$$

For the left inequality, $-3 < 2x - 3$, the solution is the set of all values for which $x > 0$. Call these values Set A.

For the other inequality, $2x - 3 \leq 5$, the solution is the set of all values for which $x \leq 4$. Call these values Set B.

For the combined inequality, the solution is

$$0 < x \leq 4$$

The intersection of Sets A and B (or A ∩ B) is pictured below:

$-3 < 2x - 3$ $2x - 3 \leq 5$ $-3 < 2x - 3 \leq 5$

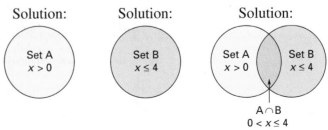

1 Solve and graph $-3 > 2x - 3$.

2 Solve and graph $2x - 3 \geq 5$.

3 What is the solution to $-3 > 2x - 3$ and $2x - 3 \geq 5$?

4 Use a Venn diagram to picture the solution to the conjunction.

5 Explain why the solution is called the *empty set*.

Disjunctions

Wire stock used to make nails is tested on a machine that compresses the wire and then stretches it. An 8-inch length of wire fails the test if it is compressed to a length shorter than 7.95 inches or stretched to a length longer than 8.05 inches. A compound inequality models the wire length that fails the test:

$x > 8.05$ *or* $x < 7.95$, where x is the length of the nail wire

This compound inequality consists of two simple inequalities connected by the word *or*. It is an example of a **disjunction.** The solution set of the disjunction is the set of all solutions that make *either* of the simple inequalities true.

EXAMPLE 2 Graphing a Disjunction

Solve and graph $-3 > 2x - 3$ or $2x - 3 \geq 5$.

SOLUTION

Solve each simple linear inequality.

$$-3 > (2x - 3) \qquad or \qquad (2x - 3) \geq 5$$

$$-3 + 3 > (2x - 3) + 3 \qquad or \qquad (2x - 3) + 3 \geq 5 + 3$$

$$0 > 2x \qquad or \qquad 2x \geq 8$$

$$0 > x \qquad or \qquad x \geq 4$$

Graph the two inequalities on a number line.

$(0 > x)$ ⟵ $(x \geq 4)$
-7 -6 -5 -4 -3 -2 -1 0 1 2 3 4 5 6 7 8 9

All the points for $0 > x$ and $x \geq 4$ constitute the solution for the combined inequality connected by the word *or*.

To check the solution to the disjunction, check x-values less than zero and x-values greater than 4 to show that the values make the combined inequality true. Then check x-values greater than or equal to zero and less than 4 to show that these values do not satisfy either inequality.

Ongoing Assessment

Solve and graph $8y + 1 < 25$ or $-2y < -10$.

The solution set of a disjunction is the **union** of two sets of numbers. The union of two sets of numbers consists of all the numbers that are in either one of the sets. The shaded area in the Venn diagram below represents the union of sets A and B. The union is designated $A \cup B$.

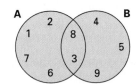

Set $A = \{1, 2, 3, 6, 7, 8\}$
Set $B = \{3, 4, 5, 8, 9\}$

Thus, the union of the two sets A and B, designated as $A \cup B$, is the set

$$A \cup B = \{1, 2, 3, 4, 5, 6, 7, 8, 9\}$$

ACTIVITY 2 Graphing a Subset

1 Solve and graph $5y - 2 < 8$.

2 Solve and graph $3y + 4 < 16$.

3 What is the solution to $5y - 2 < 8$ or $3y + 4 < 16$?

4 Use a Venn diagram to picture the solution to the disjunction.

5 Explain why the solution for $5y - 2 < 8$ is called a *subset* of the solution for $3y + 4 < 16$.

Critical Thinking In the language of algebra, AND statements correspond to the intersection of two sets, and OR statements correspond to the union of two sets. How is this similar to the way the words are used in probability?

LESSON ASSESSMENT

1 What steps are the same when you solve a conjunction and a disjunction? What steps are different?

2 How is the word AND used to find the solution of a conjunction? How is the word OR used to find the solution to a disjunction?

3 How do you know if a compound inequality is a conjunction or a disjunction?

Practice and Problem Solving

Solve each compound inequality. Graph the solution.

4. $2x \leq 6$ or $x > 5$

5. $3x < 2.4$ and $x \geq -1.5$

6. $2x - 3 < 13$ and $x > 4$

7. $5x + 4 < -11$ or $3 - 4x < -1$

8. $3x + 1 > 10$ or $2x - 5 < 3$

9. $x + 3 < -1$ and $4x - 1 < 5$

10. $-5 < 7 - 2x < 5$

11. $3 < 4x + 7 < 15$

Write and solve a compound inequality for each situation.
12. Tad will spend at least $20 but no more than $50 on T-shirts for summer camp. The cost of a T-shirt is $8. How many T-shirts can Tad buy?

A thermostat will send an electrical signal if it senses a temperature below 65 degrees or above 80 degrees.

13. If the initial temperature increases by 8 degrees, the thermostat will send a signal. What are the possible initial temperatures?

14. If the initial temperature decreases by 8 degrees, the thermostat will send a signal. What are the possible initial temperatures?

15. A store owner adds $5 to twice the wholesale price in order to set the retail price of his sweaters. If the retail prices are between $145 and $155, what wholesale prices did the store owner pay for the sweaters?

Mixed Review

The heights in inches and weights in pounds of 12 students are recorded.

Ht. 68 71 70 69 72 70 69 71 65 65 68 72
Wt. 160 165 160 140 180 155 145 160 140 135 139 150

16. Find the mode of the heights.

17. Find the median of the weights.

18. Make a scatter plot from the data. Let height be the independent variable.

19. Describe the correlation between the two sets of data.

Solve each system of equations.
20. $3a + 2b = 0$ **21.** $4c - 5d = -8$
$\quad\quad a - 3b = -7$ $\quad\quad 6c + 2d = 7$

Write and solve a system of equations that models each situation.
22. The length of the molding around the ceiling of a rectangular room is 120 feet. The length of the room is twice its width. What are the dimensions of the ceiling?

23. Mario has saved $20 less than Nancy. Together they have saved $100. How much has each person saved?

24. A chemist needs 400 grams of a 5% salt solution. There is one available solution containing 4% salt and another available solution containing 8% salt. How many grams of each solution should the chemist mix to get the desired 5% solution?

25. Mal bought 15 gallons of 87 octane gas and 12 gallons of 92 octane gas for $34.20. Yuri bought 10 gallons of 87 octane gas and 7 gallons of 92 octane gas for $21.45. How much does a gallon of each type of gas cost?

Gasoline

Plus 89   Regular 87   Premium 92

LESSON 9.5 SOLVING ABSOLUTE VALUE INEQUALITIES

The construction plans for a new bridge specify steel reinforcement rods with a diameter of 1.00 centimeter. A rod is acceptable if its diameter is not more than 0.01 centimeter from the specification; that is, the tolerance is ± 0.01 cm. An inequality containing an absolute value models the acceptable diameter of the rod.

$$|d - 1.00| \leq 0.01 \text{ where } d \text{ is the diameter of the rod in}$$
$$\text{centimeters}$$

This statement indicates that the difference between 1.00 centimeter and the actual diameter d cannot exceed 0.01 centimeter.

Recall that the definition of the absolute value of x is the distance between zero and x on a number line.

$$|x| = x, \text{ when } x \geq 0 \text{ and } |x| = -x, \text{ when } x < 0$$

By applying the definition of absolute value, you can write an absolute value equality as a disjunction. For example,

$$|x| = 2 \text{ is equivalent to the disjunction } x = 2 \text{ or } x = -2$$

The solution to an absolute value inequality such as $|x| > 2$ is a disjunction:

$$|x| > 2 \text{ is equivalent to the disjunction } x > 2 \text{ or } x < -2$$

The graph of $|x| > 2$ is shown below. Notice that all the points in the solution are a distance greater than 2 units from zero.

On the other hand, the solution to an absolute value such as $|x| < 2$, is a *conjunction*.

$$|x| \leq 2 \text{ is equivalent to the conjunction } x \leq 2 \text{ and } x \geq -2$$

The graph of $|x| \leq 2$ is shown below. Notice that all the points in the solution are a distance less than or equal to 2 units from zero.

You can also use disjunctions and conjunctions to represent absolute value inequalities that contain *linear functions*.

> **Solutions to Absolute Value Linear Functions**
> For all numbers a, b, and c, where $c \geq 0$,
>
> If $|ax + b| \geq c$, then $ax + b \geq c$ or $ax + b \leq -c$
>
> If $|ax + b| \leq c$, then $ax + b \leq c$ and $ax + b \geq -c$

Now look at the inequality for the diameter of the steel rod. Since it is an absolute value using *less than* or *equal to*, the solution must be the conjunction

$$(d - 1.00) \leq 0.01 \qquad \text{and} \qquad (d - 1.00) \geq -0.01$$

Thus,

$$(d - 1.00) + 1.00 \leq 0.01 + 1.00 \quad \text{and} \quad (d - 1.00) + 1.00 \geq -0.01 + 1.00$$
$$d \leq 1.01 \qquad \qquad \text{and} \qquad \qquad d \geq 0.99$$

Thus, an acceptable rod diameter is less than or equal to 1.01 centimeters and greater than or equal to 0.99 centimeters. In most industrial settings, this statement is written in the following way:

$$d = 1.00 \, \text{cm} \pm 0.01 \, \text{cm}$$

EXAMPLE Solving Absolute Value Inequalities

Solve $|2x - 1| > 5$.

SOLUTION

Since the absolute value is a linear inequality containing a *greater than* relation, the solution is a disjunction.

	Case 1		**Case 2**
	$2x - 1 > 5$	or	$2x - 1 < -5$
	$2x - 1 + 1 > 5 + 1$	or	$2x - 1 + 1 < -5 + 1$
	$2x > 6$	or	$2x < -4$
	$x > 3$	or	$x < -2$

The solution is the union of the two individual graphs.

Check the solution by choosing values from the three regions.

Solve and graph each absolute value inequality.

a. $|2r - 1| \geq 5$ **b.** $|5y + 2| < 3$ **c.** $|w - 4| \leq -2$

Critical Thinking For which values of a are there no solutions
for $|x - 4| < a$?

LESSON ASSESSMENT

Think and Discuss

1 Why must you be able to solve a compound inequality before
you can solve an absolute value inequality?

2 What determines whether an absolute value inequality has a
solution that is a conjunction or a disjunction? Give examples.

3 Explain why $|x + 2| \leq -4$ has no solution.

Practice and Problem Solving

Solve each absolute value inequality. Graph the solution.
4. $|x| < 5$ **5.** $|x| > 4$ **6.** $|x| \leq -1$

7. $\left|\frac{2}{3}x - 5\right| \geq 3$ **8.** $|x - 2| < \frac{1}{3}$ **9.** $|4x + 5| > 15$

10. $|x - 1| > 7$ **11.** $|5 - 3x| \geq 10$ **12.** $|3x - 4| \leq 8$

Write and solve an absolute value inequality that models
each situation.

13. The arrival of an airline
flight was predicted to be
within 5 minutes of its
scheduled flight time of 120
minutes. If the flight did not
arrive within its predicted
time, what was the airline's
actual flight time?

14. A survey by a marketing research company shows that
56% of the customers polled prefer Super Cola over all other
soft drinks. The survey has a margin of error of 13%. What
percent of all soft drink customers prefer Super Cola?

15. The warranty for Captain Tires recommends a pressure of 30 pounds per square inch (psi). A tire is considered unsafe if its pressure is more than 3 psi from the recommended pressure. What is the unsafe pressure range for Captain Tires?

16. An aircraft design specifies a structural support beam as 1.2 centimeters thick, with an allowable error in thickness of 0.04 centimeters. If you are fabricating this beam, what is the allowable thickness of your final product?

Mixed Review

A warehouse needs exhaust fans. The warehouse is 500 feet wide and 700 feet long. It has a ceiling that is 25 feet high. Each fan moves 500 cubic feet of air per minute. The air in the warehouse must be replaced three times per hour.

17. What is the volume of the warehouse in cubic feet? Express your answer in scientific notation.

18. What volume of air is replaced in the warehouse each hour? Each minute? Express your answers in scientific notation.

19. How many exhaust fans are required?

20. The expression 16a represents the perimeter of a square. Write the expression that represents the area of the square.

Write an equation in one unknown to model each situation. Solve the equation.

21. Tickets to the theater are $9. If 25 people buy tickets as a group, there is a 25% discount. How much is one ticket at the discounted price?

22. You can buy six greeting cards for $10. How much will it cost to buy nine cards?

23. The sum of three consecutive numbers is 63. What are the numbers?

24. The Book Club sells books for the wholesale cost plus $3.50. The Book Club receives a check for $151 to purchase five books. What is the wholesale cost of one book?

LESSON 9.6 GRAPHING LINEAR INEQUALITIES IN TWO VARIABLES

You learned to solve and graph linear equations on a coordinate plane in Chapter 8. Solving and graphing inequalities with two variables is very similar. You can graph linear equations or inequalities on a coordinate plane with one of these methods:

- *Use a table to list ordered pairs.*
- *Use the slope-intercept form of the equation.*
- *Use a graphics calculator.*

Consider the following equation and inequality:

$$4x + 2y = 6 \qquad 4x + 2y \le 6$$

Solve for y and write both the equation and the inequality in slope-intercept form ($y = mx + b$ and $y \le mx + b$).

$$4x + 2y = 6 \qquad\qquad 4x + 2y \le 6$$

$$2y = -4x + 6 \qquad\qquad 2y \le -4x + 6$$

$$y = -2x + 3 \qquad\qquad y \le -2x + 3$$

Now graph the equation and the inequality.

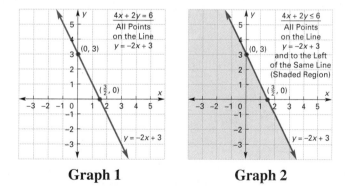

Graph 1   **Graph 2**

Look at Graph 1. If you pick a point on the line and substitute this ordered pair into $y = -2x + 3$, the result is a true statement.

The same is true for Graph 2. However, the graph of the inequality contains both the line $y = -2x + 3$ and the shaded area to the left of the line. The shaded area is the graph of $4x + 2y < 6$. Together, the line and the shaded area are the graph of $4x + 2y \leq 6$.

Critical Thinking   Describe the graph of $4x + 2y \leq 6$ as the union of two sets.

If you choose a point that is on the line or in the shaded area and substitute this ordered pair into $4x + 2y \leq 6$, the result will be a true statement.

Use the ordered pair $(4, -5)$, which is on the line.

$$4x + 2y = 6$$

$$4(4) + 2(-5) = 6$$

$$6 = 6$$

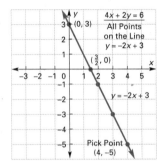

The ordered pair makes the statement true.

Now use $(-1, 2)$ in the shaded region and $(2, -1)$ on the line.

Shaded Region: $(-1, 2)$	On the Line: $(2, -1)$
$4x + 2y \leq 6$	$4x + 2y \leq 6$
$4(-1) + 2(2) \leq 6$	$4(2) + 2(-1) \leq 6$
$-4 + 4 \leq 6$	$8 - 2 \leq 6$
$0 \leq 6$	$6 \leq 6$

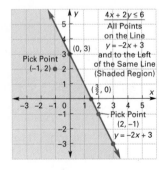

Each ordered pair results in a true statement.

To graph a linear inequality with two variables, first graph the line that represents the associated equation. The graph of the equation divides the coordinate plane into three parts:

- *the line itself*
- *the area to one side of the line*
- *the area to the other side of the line*

EXAMPLE 1 Describing Graphs of Inequalities

Write an equation or inequality to describe the following graphs. Each of these graphs compares the expression $(-2x + y)$ to 5 using an equality or an inequality.

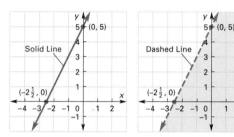

SOLUTION

a. $-2x + y = 5$
$y = 2x + 5$

b. $-2x + y < 5$
$y < 2x + 5$

c. $-2x + y > 5$
$y > 2x + 5$

Notice that in the graphs for inequalities, the lines are dashed. This means that the line itself is not included in the solution set. If the inequalities use $\leq$ or $\geq$, use a solid line to show the lines are included in the solution set.

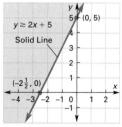

To determine which side of the line to shade, choose a *test point* above or below the line and substitute the corresponding ordered pair into the inequality. (Use the test point $(0, 0)$ whenever you can.) Check to see if the result is a true statement. If it is, you have chosen the side that represents the solution. Shade this side of the line.

EXAMPLE 2 Choosing the Area to Graph

Graph the inequality $-x + y \leq 2$.

SOLUTION

Use a table or change the inequality to slope-intercept form, $y \leq x + 2$. Then draw the graph.

First draw the graph of the line $-x + y = 2$. Then substitute test points on either side of the line to determine which side to shade.

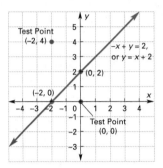

Above the Line: $(-2, 4)$	Below the Line: $(0, 0)$
$-x + y \leq 2$	$-x + y \leq 2$
$-(-2) + 4 \leq 2$	$0 + 0 \leq 2$
$6 \leq 2$	$0 \leq 2$

This is false. Thus, $(-2, 4)$ is not a solution. Do not shade above the line. This is true. Thus, $(0, 0)$ is a solution. Shade below the line.

Look again at Example 2. If you change an inequality to slope-intercept form, you can tell by inspection whether you should shade above or below the line. If the inequality is $<$ (less than), shade below the line. If the inequality is $>$ (greater than), shade above the line.

Ongoing Assessment

Graph each inequality.

a. Use a test point to determine the region to shade.

   1. $(-4x + y) < -1$ 2. $(-4x + y) \geq -1$

b. Then change each inequality to slope-intercept form and use inspection to determine whether to shade above or below the line.

The following examples illustrate how inequalities help solve problems that arise in the workplace.

EXAMPLE 3 A Profit Model

Suppose Mountain Sales Bicycle Shop makes $100 on each Model X bike sold and $50 on each Model Y bike sold. The bike shop's overhead expenses are $1500 per month. At least how many of each model bike must be sold each month to avoid losing money?

SOLUTION

The bike shop must make at least $1500 each month. The condition "at least $1500" is represented by "≥ 1500." The "equal" part of $\geq$ means that if the owner makes exactly $1500, he meets expenses. The "greater than" part of $\geq$ means that if the owner makes over $1500, he meets expenses and makes a profit.

Let x and y represent the number of Model X and Model Y bikes sold per month. Since the owner must make at least $1500, an inequality models the situation.

$$100x + 50y \geq 1500$$

Solve for y to write the inequality in slope-intercept form.

$$100x + 50y \geq 1500$$

$$50y \geq -100x + 1500$$

$$y \geq -2x + 30$$

The line that defines the boundary of the shaded region is $y = -2x + 30$. Since the owner cannot sell less than zero of either model bike, the graph is confined to the first quadrant, where x- and y-values are both greater than or equal to zero.

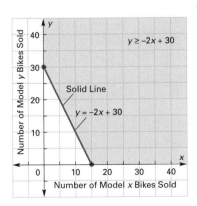

Conclusions

1. All points in the shaded region, on and above the line, are solutions.

2. All points below the line (unshaded triangle) are not solutions.

Ongoing Assessment

Determine whether the following points are in the solution set and, if so, how much money the bike shop makes.

$$(7, 12); \quad (10, 10); \quad (15, 0); \quad (0, 29); \quad (20, 20)$$

EXAMPLE 4 Graphing Distance Versus Time

A truck is traveling on a road with a posted speed limit of 55 miles per hour. An inequality that describes the distance that could be covered by a driver operating the truck legally is $d \leq 55t$. In this inequality, d is the distance the truck travels and t is the time it takes to travel that distance.

SOLUTION

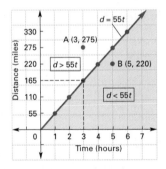

The lower shaded region represents the legal speed zone, where $d < 55t$; that is, d is less than $55t$ for any pair (t, d) in this region.

The upper shaded region represents the illegal speed zone, where $d > 55t$; that is, d is greater than $55t$ for any pair (t, d) in this region.

Since the point $(3, 165)$ lies on the line, the coordinates satisfy the equation $d = 55t$. What does the number 165 represent?

Point A has coordinates $(3, 275)$. Point A is "typical" of all points in the region above the line $d = 55t$. At what speed would a truck be traveling to satisfy coordinates $(3, 275)$? Is this a legal speed in the speed zone of Example 4?

Point B has coordinates (5, 220). Point B is "typical" of all points in the region below the line $d = 55t$. At what speed would a truck be traveling to satisfy coordinates (5, 220)? Is this a legal speed in the speed zone of Example 4?

EXAMPLE 5 Multiplying by a Negative Number

Graph this inequality.

$$-3x - 2y > -2$$

SOLUTION

First, write the inequality in slope-intercept form.

$$-3x - 2y > -2$$

$$-2y > 3x - 2$$

$$y < -\frac{3}{2}x + 1 \quad \text{(Note that} > \text{changes to} < \text{after division by} -2.)$$

Next, graph $y = -\frac{3}{2}x + 1$ as a dashed line. Then determine the shaded region.

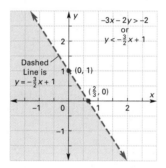

Since the inequality symbol in the slope-intercept form is $<$, the shaded area must be below the dashed line, as shown.

Choose a test point in the shaded region of the graph and substitute this ordered pair into both forms of the inequalities:

$$-3x - 2y > -2 \quad \text{and} \quad y < -\frac{3}{2}x + 1$$

Are the inequalities true?

$-3 - 2y > -2$	$y < -\frac{3}{2}x + 1$
choose (0,0)	choose (0,0)
$-3 > -2$	$0 < 1$

LESSON ASSESSMENT

Think and Discuss

1 How do you determine if the line associated with an inequality is graphed as a dashed or a solid line?

2 How do you determine which side of the line to shade when you graph an inequality?

3 Give examples of different methods for determining which side of the line graphed contains all the solutions to an inequality.

Practice and Problem Solving

Solve each inequality for y. Graph each inequality.

4. $x + y > 6$ **5.** $x + y > -2$

6. $x - y < 6$ **7.** $2x - y \geq 5$

8. $3y + 12 < 0$ **9.** $3x - 4y \leq 12$

10. $5x + 3y < 15$ **11.** $5y - 40 < -8x$

12. $6x - 3y + 12 < 0$

The perimeter of a rectangular garden is more than 24 meters.

13. Write an inequality to model the perimeter as a function of length and width.

14. Solve the inequality and graph the solution.

15. Give three whole number solutions to the inequality.

This semester, there are fewer than 30 students on the chess team.

16. Write an inequality to model the number of boys and girls on the team.

17. Solve the inequality and graph the solution.

18. If there are more girls than boys on the team, give three possible combinations of numbers of boys and girls on the team.

A manufacturing company uses two machines to cut patterns for wet suits. It costs $10 per wet suit to use the faster machine. It costs $6 per wet suit to use the slower machine. The daily cost of cutting out wet suits must be held to a maximum of $60.

19. Write an inequality to model the daily cost of cutting out wet suits as a function of the number of wet suits cut on each machine.

20. Solve the inequality and graph the solution.

21. Give three whole number solutions to the inequality.

22. A work order for one day requests 8 wet suits. How many wet suits can the company cut out on each machine and still meet the cost constraints for the day? Give two other possible answers.

Mixed Review

Suppose line *AB* passes through *A*(0, 9) and *B*(9, 0). Suppose line *MN* passes through *M*(1, 4) and *N*(−1, 2).

23. What is the slope of each line?

24. What is the equation of line *AB*?

25. What is the equation of line *MN*?

26. What is the *y*-intercept of each line?

27. Graph each line on the same coordinate axes.

28. At what point do the lines intersect?

29. Find the equation of a line *CD* that passes through the origin and is parallel to line *AB*.

30. At what point do lines *CD* and *MN* intersect?

LESSON 9.7 SOLVING SYSTEMS OF INEQUALITIES

Recall the car dealer from Chapter 8. Suppose the car dealer nets $500 for each family car (F) sold and $750 for each sports car (S) sold. The dealer must sell at least two family cars for every sports car and must earn at least $3500 per week. You can find S and F by graphing the two inequalities that meet these conditions.

Inequality	Meaning
$F \geq 2S$	sell at least 2 family cars for each sports car
$500F + 750S \geq 3500$	make at least $3,500 per week if net for each family car sold is $500 and net for each sports car sold is $750

The solution of this system of inequalities is the intersection of the solution regions of the two inequalities.

ACTIVITY Intersecting Solution Sets

1 Graph the lines $F = 2S$ and $500F + 750S = 3500$ on the same axes.

2 Use test points to determine the solution region for each inequality.

3 Find the area where the two solution regions intersect. This is the solution region for the system of inequalities.

A solution region contains many solutions. In the car dealer problem, the dealer would want to know which solution maximizes profit. You can use linear programming to determine the *optimum* number of items to sell to *maximize* profits.

Recall the Mountain Sales Bicycle Shop problem. The shop made $100 on each Model X bike sold and $50 on each Model Y bike sold. The owner's objective is to maximize sales. You can model sales with the following equation:

$$Sales = 100x + 50y$$

In linear programming, this equation is called an **objective function**, because the objective is to maximize sales. In the objective function, x is the number of Model X bikes sold and y is the number of Model Y sold. Suppose that each month the owner can have in stock no more than 20 Model X bikes and no more than 30 Model Y bikes. These conditions are represented by inequalities.

$$x \leq 20 \quad \text{and} \quad y \leq 30$$

Furthermore, suppose the maximum number of bikes the shop can stock is 40. This condition is written as an inequality.

$$x + y \leq 40$$

Since the owner cannot order fewer than zero of either model, a final condition also exists.

$$x \geq 0 \quad \text{and} \quad y \geq 0$$

In the language of linear programming, each of the five inequalities listed above is called a **constraint**. Each inequality limits or constrains the set of values for x and y. This is illustrated in a graphical representation of the problem.

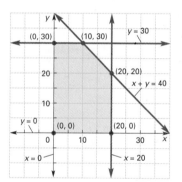

The shaded region is the intersection of all the constraining inequalities. The shaded region is a polygon bounded by the five lines that represent the constraints. All **feasible solutions** to the problem are contained in the shaded region. Note particularly the five corner points around the perimeter of the polygon: (0, 0); (20, 0); (20, 20); (10, 30); and (0, 30). To find the **optimum solution**, substitute each of these five coordinate pairs into the objective function (sales equation) and compare the sales.

Feasible Solution	Objective Function (Sales)
0, 0	100(0) + 50(0) = 0
20, 0	100(20) + 50(0) = 2000
20, 20	100(20) + 50(20) = 3000
10, 30	100(10) + 50(30) = 2500
0, 30	100(0) + 50(30) = 1500

The table shows that the maximum sales for the bike shop is $3000. This occurs when the bike shop stocks and sells 20 of each model each month. Although any point in the shaded region of the graph of the system of inequalities satisfies all five inequalities (constraints), only the corner point (20, 20) maximizes the sales.

WORKPLACE COMMUNICATION

Check Mr. Cannon's math. Let *x* and *y* be the number of miles driven each day. Write an inequality modeling rule 5(a). Write two more inequalities involving *x* and *y* for rule 5(c). Graph the three inequalities to determine the combinations of *x* and *y* that satisfy both safety rules. If the distance between MWI and Carlisle is 775 miles, can the trip be made in two days within the safety rules?

Jackson Trucking Company
1-400-569-4398

Modern Workplace, Inc.
Attn: Ms. Anita Villarreal
          Chief, Environmental Control

Dear Ms. Villarreal:

I am writing in response to your request for clarification concerning our proposal to transport hazardous waste from the MWI facility to the Carlisle waste treatment plant.

Our cost proposal for this job is based on using a single driver for two days for each trip. Our internal safety rules, approved by state and federal transportation authorities, allow a single driver for hazardous cargo, subject to layover restrictions. The restrictions are described in the Jackson Trucking Safety Rules. I have attached a copy of the relevant page for your review.

I hope this clarifies our proposed use of a single driver. If you have any further questions, please do not hesitate to call me at 232-1600. Thank you for considering our proposal. I am sure Jackson Trucking can satisfy all MWI's hazardous material transportation needs.

Sincerely,

Graham Cannon

President

Attachment

JACKSON TRUCKING CO.
Hazardous Materials On Board

state regulations concerning transportation of hazardous cargo.

5. Hazardous cargo may be transported by a single driver; however, in such cases:
   (a) The average distance driven per day shall be less than 400 miles.
   (b) Trips greater than 400 miles require layovers. The driver shall rest at least eight hours during a layover.
   (c) No distance driven in one day shall be greater than 475 miles.

LESSON ASSESSMENT

Think and Discuss

1 What conditions must be true for a system of inequalities to have a common solution?

2 If the lines for two inequalities are parallel, can there be a common solution? Why or why not?

Graph each system of inequalities. Shade the area to show the solution.

3. $2x + y \geq 4$
$y - 2x \geq 4$

4. $x - y > -1$
$x - y \leq 2$

5. $2x + y \leq 4$
$y - 2x \leq 4$

6. $5x + 3y \leq 1$
$3x - 2y > 12$

7. $2x + 5y < 4$
$3x + 4y > -1$

8. $5x - 7y > 4$
$2x + 3y < 18$

9. $3x - 2y < 2$
$4y - 3x > 2$

10. $2y - 6x \leq 3$
$3y + 5x < -6$

11. $5y + 3x \leq 0$
$3y - 2x \leq 0$

A painting contractor estimates it will take 10 hours to paint a one-story house and 20 hours to paint a two-story house. The contractor submits a bid to paint 20 houses in less than 250 hours.

12. Write a system of equations to model the time to paint the houses and the number of houses to be painted.

13. Graph the system.

14. Give five whole-number solutions to the system. Explain what these solutions mean.

It costs 50 cents to make a bracelet and $1 to make a necklace. To make a profit, the total cost for bracelets and necklaces must be less than $10. The jeweler can make no more than 14 pieces of jewelry each day.

15. Write a system of inequalities to model the number of bracelets and necklaces to be made each day.

16. Graph the system.

17. The jeweler sells bracelets for $3 and necklaces for $4. Test the three corner points on your graph and determine how many bracelets and necklaces should be made to maximize profits.

Write a system of inequalities for each situation. Graph the solution.

18. A farmer needs to enclose a rectangular field with wire fencing. He has 800 meters of fencing available. The length of the field must be greater than 100 meters. What are the possible dimensions of the field? What is the maximum possible area of the field?

19. A manufacturer makes running shoes and basketball shoes. It costs $20 to make a pair of running shoes and $30 to make a pair of basketball shoes. The manufacturer can spend no more than $2000 making both types of shoes each day. The company needs to make no fewer than 50 pairs of running shoes each day. How many of each type can be produced each day?

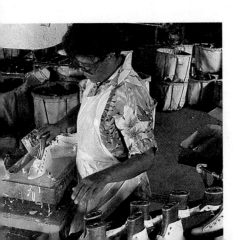

20. A hardware distributor wants to sell at least 45 electric and gas mowers each week. The profit on an electric mower is $50 and the profit on a gas mower is $40. Which combinations will give the distributor a profit of up to $2500 per week?

Mixed Review

A jar contains 10 nickels, 15 dimes, and 25 quarters. One coin is removed at random.

21. What is the probability that the coin is a dime?

22. What is the probability that the coin is a nickel or a quarter?

A jar contains 10 nickels, 15 dimes, and 25 quarters. Two coins are removed at random.

23. What is the probability that both coins are dimes?

24. What is the probability that one coin is a nickel and the other is a quarter?

A jar contains 10 nickels, 15 dimes, and 25 quarters. One coin is removed at random. It is replaced, and another coin is removed at random.

25. What is the probability that both coins are the same?

26. What is the probability that both coins have a sum less than 25¢?

Solve using an equation in one unknown.

27. Tim and Shaneka have together invested $98 in an after-school business. Shaneka has invested $14 more than Tim. How much does each person have invested?

Solve using a system of linear equations.

28. Sherrill has $8000 invested in CD accounts. One account pays 6% simple interest, and the other pays 8% simple interest. This year Sherrill earned $580 interest on her two accounts. How much is invested in each account?

MATH LAB

Activity 1: Distribution of Student Heights

Equipment Calculator
Tape measure with metric scale

Problem Statement

You will plot the heights of the students in your class on a number line. Then you will make inequality comparisons with the height measurements of the group.

Procedure

a Use the tape measure to determine the height of each member of your group. Record the measurements to the nearest centimeter.

b Create a number line that includes the height measurements for each member of your group. Plot the height measurements on your number line. Place each member's name above the point representing his or her height. An example is shown below.

If two members have the same height, "stack" the points and their names above each other.

c Use the three relationships (<, =, and >) to write an expression relating various pairs of people. Use each group member at least once. For example, using the sample number line data above, you can write "Liz > Judy."

d Select three people from your number line and write a compound inequality. For example, using John, Pat, and Bob from the sample number line, you can write "Bob < John < Pat."

e Record your data on a master number line that includes the results for your entire class.

f Determine how many students from your class would represent 20% of all the members in the class. Round your answer to the nearest whole number. Use this number to identify the students who are in the highest 20% and lowest 20% of the class according to height. The remaining students are in the middle 60%. Write a compound inequality that describes the range of heights for the middle 60% of your class. For example, let *H* represent the height of a student in the middle 60%. Then write a compound inequality such as the following:

$$152 \text{ cm} < H < 166 \text{ cm}$$

g Can you use your inequality to predict the heights of other students in the same age group in your school? Explain your answer.

Activity 2: Sampling the Quality of a Product

Equipment Calculator with statistical functions
20 large machine bolts and nuts
Vernier caliper
Micrometer caliper

Problem Statement

Most products include tolerances in their specifications. In this laboratory, you will measure the dimensions of a sample of machine bolts. Then, based on your measurements, you will determine how well the machine bolts meet a given set of tolerances.

Procedure

a Measure the length of each of the 20 bolts with the vernier caliper, and measure the head width of each bolt with the micrometer caliper. Record the results in a table similar to the one shown below:

Head Width	Overall Length
0.789 in.	5.11 cm
.	.
.	.
.	.
Mean: _____	Mean: _____

b Write six inequality statements that describe the maximum and minimum measured values. The six statements will include:

Two *greater than* relationships. For example, head width > 0.760 in., length > 5.03 cm

Two *less than* relationships. For example, head width <0.792 in., length < 5.16 cm

Two *compound inequality* relationships. For example, 0.760 in. < head width < 0.792 in.; 5.03 cm < length < 5.16 cm

c Create a histogram of the head width measurements and also of the length measurements. Adjust the width of each cell so that the data are spread over six to eight cells. For example, suppose your range of measurements for head width was 0.032 inches—that is, from 0.760 inches to 0.792 inches—Then the width of each cell could be $\frac{0.032}{8} = 0.004$, and your histogram might look like the sample.

d Use your calculator to determine the mean of the head width measurements and the mean of the length measurements. Use the statistics keys or the formula for the mean. Record the mean for each set of data in your table, under the appropriate column.

e If you work in quality assurance for a company, you may help establish tolerances for a production process. Quality assurance personnel sample the product under ideal production conditions and recommend a tolerance for the particular machine or process. For this activity, use the following tolerance for the head width.

$$\overline{w} \text{ inches} \pm 0.003 \text{ inches}$$

where $\overline{w}$ is the mean head width from Step **d**. Use the following tolerence for the length.

$$\overline{L} \text{ cm} \pm 0.01 \text{ cm}$$

where $\overline{L}$ is the mean length from Step **d**.

f Write the tolerances in Step **e** as compound inequalities. Use the horizontal axes of your histograms as number lines. Graph your inequalities on the appropriate number line.

g Do any of your bolt measurements fall outside the tolerance you wrote in Step **f**?

h Compare your inequalities with those of the other groups in your class. Do any of their bolt measurements fall outside your group's tolerance interval?

Activity 3: Linear Programming

Equipment   Calculator
Paper
Pencil
Scissors
Stopwatch

Problem Statement

Production line schedules are determined in advance to maximize a manufacturer's profits. Among the many factors that determine production schedules are: profit per item, time to produce the item, the total number of items to produce, and available worker hours.

In this Activity, you will measure the average time it takes to produce two different production items. You will then use this data with given information to graph inequalities. Finally, you will use a form of linear programming to determine how many of each item should be produced to maximize profits.

Use the following information to complete the Activity:

x is the total number of Widget 1s.
y is the total number of Widget 2s.
The total time available to construct all of the widgets cannot exceed 30 minutes (1800 seconds).
The total number of widgets constructed will not exceed 35.
The profit from the sale of one Widget 1 is $0.75.
The profit from the sale of one Widget 2 is $1.50.

Procedure

a Have one student in the group trace five outlines of Widget 1, shown below. Trace all lines and the center dot. Have a second student use a stopwatch to time the first student in completing the tracing of the five outlines of Widget 1. Start the stopwatch when the pencil first touches the paper. Stop the watch when the pencil is lifted from the paper after the fifth Widget 1 is outlined. Record the time in seconds under the column "Total Tracing Time" in the data table shown below.

Object	Total Tracing Time (sec)		Average Tracing Time (sec)	Total Cutting Time (sec)		Average Cutting Time (sec)
Widget 1		÷5			÷5	
Widget 2		÷5			÷5	

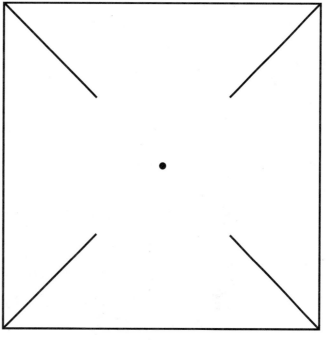

Widget 1

b Have a third member of the group cut out the five outlines of Widget 1. Make certain to cut the length of each heavy line toward the center. The second student will time the cutting of the five outlines of Widget 1. Start the stopwatch when the scissors first touch the paper. Stop the watch when the cutting of the fifth Widget 1 is completed. Record the time in seconds under the column "Total Cutting Time" in the Data Table.

c Repeat Steps **a** and **b** for Widget 2 shown below.

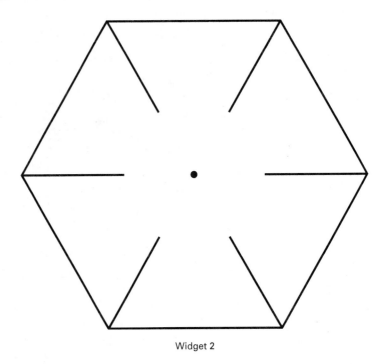

Widget 2

d Calculate the average time to trace one Widget 1. Round your answer *up* to the next whole second. Record this time in the Data Table under the column "Average Tracing Time." Repeat the process for Widget 2.

e Calculate the average time to complete the cutting of one Widget 1. Round your answer *up* to the next whole second. Record this time in the Data Table under the column "Average Cutting Time." Repeat the process for Widget 2.

f Use the information in the Problem Statement to write an equation for the profit P in selling x Widget 1s and y Widget 2s.

g Write inequalities that show the following:

- You cannot make fewer than zero of either type of widget. (Note: There will be one inequality for x and one for y.)

- The total time to trace and cut out x Widget 1s cannot exceed 30 minutes. Use the average times for tracing and cutting found in Steps **d** and **e** to write this inequality.

- The total time to trace and cut out y Widget 2s cannot exceed 30 minutes (1800 seconds). Use the average times for tracing and cutting found in Steps **d** and **e** to write this inequality.

- The total number of widgets constructed cannot exceed 35.

These five inequalities are the constraints that limit the profit from widget production.

h Graph the five inequalities from Step **g** on a common set of axes.

Calculations

1. Determine the coordinates of the points of intersection of the lines that represent the constraints. Record the coordinates in the table below.

2. Evaluate the equation in Step **f** using the coordinates above to determine the profit for this combination of sales. Record the profit in the table.

Coordinates	Profit
(,)	
(,)	
(,)	
(,)	
(,)	

3. How many of each type of widget should be made to produce the maximum profit?

MATH APPLICATIONS

The applications that follow are like the ones you will encounter in many workplaces. Use the mathematics you have learned in this chapter to solve the problems.

Wherever possible, use your calculator to solve the problems that require numerical answers.

Rewrite each sentence as an inequality. For example, "Valarie received a lower score on the algebra test than Vicky" can be written as

Valarie's score on the algebra test < Vicky's score on the algebra test.

1 Bill is taller than Jack but not as tall as Sam.

2 This car will cost at least $10,500.

3 The most I can spend for a new suit is $160.

4 Phil is the same age as Janet, but he is older than Richard.

5 The water level does not reach the top of the dam.

6 The trip will be canceled unless at least 12 persons are scheduled to go.

Match each inequality with the correct number line in the list at the right.

7 $n \leq 0$

8 $n > 0$

9 $n = 0$

10 $n < 5$

11 $n \geq 5$

12 $5 < n$

13 $0 \leq n \leq 5$

P: −2 −1 0 1 2 3 4 5 6 7

Q: −2 −1 0 1 2 3 4 5 6 7

R: −2 −1 0 1 2 3 4 5 6 7

S: −2 −1 0 1 2 3 4 5 6 7

T: −2 −1 0 1 2 3 4 5 6 7

U: −2 −1 0 1 2 3 4 5 6 7

V: −2 −1 0 1 2 3 4 5 6 7

W: −2 −1 0 1 2 3 4 5 6 7

The lengths (in inches) of several fish drawn from a lake on two separate occasions are listed as Set A and Set B.

Set A = {8, 9, 10, 11, 12}
Set B = {10, 11, 12, 13, 14}

14 Write the union of Sets A and B: A ∪ B.

15 Write the intersection of Sets A and B: A ∩ B.

16 If the length of a fish in Set A is represented by *f*, write an inequality that describes the interval of Set A.

17 Let *g* represent the length of a fish in Set B. Write an inequality for the interval of Set B.

18 Let *h* represent a number of the intersection of Set A and Set B. Write an inequality for the interval of the intersection.

In Exercises 14 and 15, you found the union and the intersection of two sets, A and B. Suppose you obtain a third sample of fish lengths, Set C.

Set A = {8, 9, 10, 11, 12}

Set B = {10, 11, 12, 13, 14}

Set C = {8, 11, 12, 7, 15}

19 Use the results of Exercise 14 to find the union of Sets A and B with Set C: (A ∪ B) ∪ C.

20 Find the union of Sets B and C: B ∪ C. Use the results to find the union of Set A with the union of Sets B and C: A ∪ (B ∪ C). Compare your answer to the results of Exercise 19. Do you think the following equality is true for any Set A, B, and C?

(A ∪ B) ∪ C = A ∪ (B ∪ C)

21 Use Sets A, B, and C to see if the following equality is true.

(A ∩ B) ∩ C = A ∩ (B ∩ C)

You need a semester average of at least 80 in Algebra to stay on the track team. You have taken six tests but you must still take the semester exam. The semester exam counts as two test grades when computing the semester average. Your six test grades are 88, 77, 75, 80, 80, and 70.

22 Write a formula to determine your semester test average *T*, if *E* is your score on the semester exam.

23 Write an inequality for the semester test average that describes an average of at least 80.

24 Combine the results of Exercises 22 and 23 to write an inequality that contains one variable, *E*.

25 Solve the inequality to find the score that you must make on the semester exam to stay on the track team.

AGRICULTURE & AGRIBUSINESS

Your landscaping company has been hired to mow a field with an area of 150,000 square yards. You have two mowers. To finish this job in 8 hours, you must rent additional mowers. You must rent no more mowers than needed. Each mower can cut at a rate of 2500 square yards per hour.

26 Determine how many square yards each mower can cut in 8 hours.

27 Write an expression for the area that can be mowed by your two mowers and *n* additional mowers (in 8 hours).

28 Write an inequality that relates the expression from Exercise 27 to the total mowing area.

29 Solve your inequality for *n* to find the fewest number of mowers you should rent to finish the job in eight hours.

Mr. Jackson owns 2800 acres of farmland. He must decide how much of two different types of crops to plant. When Mr. Jackson sells Crop A, he expects to make $240 per acre. When he sells Crop B, he expects to make $270 per acre. Federal regulations limit Mr. Jackson's planting to no more than 2000 acres of Crop A and no more than 1200 acres of Crop B.

30 Mr. Jackson plants *a* acres of Crop A and *b* acres of Crop B. Write an inequality for the total number of acres Mr. Jackson can plant.

31 Write inequalities for the total number of acres that can be planted for each type of crop.

32 Construct a graph for *a* versus *b*. Graph all the inequalities and shade the area that contains the points which satisfy all the conditions of your inequalities; that is, shade the region that contains feasible solutions for Mr. Jackson.

33 Mr. Jackson wants to maximize income from the sales of his crops. Write an equation for the amount of income (I) from the sale of Crop A and Crop B.

34 Using the techniques of linear programming, find the number of acres of Crop A and Crop B that satisfies the constraints and results in the maximum income.

BUSINESS & MARKETING

A small accounting firm wants to subscribe to a nationwide database service. The service charges a monthly fee of $50 plus a $15 per hour access fee. The firm does not want to spend more than $500 per month for the database service.

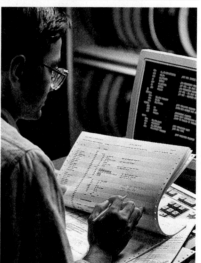

35 Write an inequality that describes the relationship between the maximum amount the firm wants to spend and the total cost per month.

36 Solve the inequality to find the number of hours the accounting firm can be connected to the database and stay below budget.

As a travel agent, you have been hired to schedule a trip for a school's French club. The club has raised $5500 for the trip. Your agency charges a setup fee of $250. The cost per person will be $500. You must keep the cost below $5500.

37 Write an inequality that shows the relationship between the travel agency charges and the amount the club has raised.

38 Solve the inequality to find the number of club members that can go on the trip within the cost constraints.

Your ice cream parlor is running a promotional sale. For every regularly priced ice cream cone, a customer can buy a sundae on sale. On every sundae you sell at the sale price, you lose $0.19. You can recover your losses with the $0.24 profit you make by selling the regularly priced cones.

39 If "breaking even" is the worst you are willing to accept from the promotion, you will want the losses from the sales of sundaes to be less than or equal to the profits from the sale of cones. Write an inequality for this situation, letting s represent the number of sundaes sold and c the number of cones sold.

40 Graph the inequality, plotting the number of sundaes sold on the horizontal axis and the number of cones sold on the vertical axis. Indicate the region where your store will profit from the promotion.

Your company produces electronic games. To meet increased demand, you must purchase a number of new circuit board assembly machines. Brand X machines can assemble 280 games per month and cost $4000 each. Brand Y machines can assemble 320 games per month, and cost $7000 each. The company must maintain a production rate of at least 4100 games per month, and your budget for purchasing the new machines cannot exceed $75,000.

41 Write an inequality that relates the total cost of buying x Brand X machines and y Brand Y machines to your maximum purchase price.

42 Write an inequality that relates the total number of electronic games that you can produce with the new machines to the minimum acceptable production rate.

43 Graph each inequality and find the numbers of each brand of machine that you could order to meet the production requirement and stay within your budget. Remember that you must order whole machines, not fractional parts of a machine.

Your computer store sells two types of computers. The profit on the sale of a Model X is $250, while the profit from the sale of the more powerful Model Y is $350.

44 Write an expression that shows the income from selling x Model X computers and y Model Y computers during a given month.

45 You want the income from the sales of these two models to be at least $2000. Write an inequality for this condition.

46 Write the inequality in slope-intercept form and draw a graph of the inequality.

47 Use your graph to list three combinations of sales of Model X and Model Y computers that would allow you to meet your income goal.

48 Use your graph to list three combinations of sales of Model X and Model Y computers that would *not* allow you to meet your income goal.

You can spend $2400 on a vacation at a health and fitness resort. The round-trip air fare is $900. Each day at the resort costs $198. You want to determine the maximum number of days you can stay at the resort.

49 Write an expression for the total cost of airfare plus staying d days at the resort.

50 Use the result of Exercise 49 to write an inequality relating the total cost of the vacation to the spending limit.

51 Solve the inequality for d to find the maximum number of days you can stay at the resort.

Robert works in the Health Physics department at a nuclear waste disposal site. His job is to monitor the radiation exposure of the employees. The National Council on Radiation Protection (NCRP) recommends that a worker receive no more than 500 millirem of radiation exposure in a one-year period. At this disposal site, Robert has measured an average dose of 1.85 millirem per day for the workers. (assuming 8-hour work days)

BIO-MEDICAL
MATERIAL

52 Write an expression for the total amount of radiation a worker receives (on the average) after working d days in a year.

53 Write an inequality relating the NCRP recommendation to the total radiation received by an employee who works d days in a year.

54 Solve the inequality to determine the maximum number of days employees should work in a year.

55 Write a sentence summarizing the result from Exercise 54.

A clothing store manager wants to restock the men's department with two types of suits. A Type X suit costs $250. A Type Y suit costs $325. The store manager needs to stock at least $7,000 worth of suits to be competitive with other stores, but the store's purchasing budget cannot exceed $10,000.

56 Write an expression for the purchase cost of Type X suits and Type Y suits.

57 Write an inequality relating the purchase cost and the minimum value of suits in stock.

58 Write another inequality relating purchase cost and the maximum purchasing budget.

59 Graph the inequalities from Exercises 57 and 58 on the same axes.

60 Find the region that satisfies both inequalities and shade this region. Name one combination of purchases that will not exceed the maximum budget, yet will provide more than the minimum inventory.

Martin manages a beauty salon that offers a facial with manicure and a haircut with wash and styling. The salon staff can do as many as 10 facials a day, or they can give up to 24 haircuts a day. If the staff does both procedures at the same time, they can use the waiting time during the procedures more efficiently. In such a case, they can perform up to 29 facials and haircuts combined. The income from the facial procedure is $55, while the haircut brings in $35.

61 Let y represent the number of facials given in a day. Write an inequality for the number of facials the salon staff can give in one day.

62 Let x represent the number of haircuts given in a day. Write an inequality for the number of haircuts that can be given in one day by the salon staff.

63 Write an inequality for the total number of haircuts and facials the staff can do when they give both at the same time.

64 Draw a graph of these three inequalities. Shade the region where all three conditions are satisfied.

65 Write a formula for the total income from x haircuts and y facials given in a day.

66 Use the techniques of linear programming to determine the optimum number of haircuts and facials the salon should give to maximize income.

A chemical formula calls for a solution of pH level 6.3 with a tolerance of ±0.4.

INDUSTRIAL
TECHNOLOGY

67 Let x represent the pH level of the solution. Use an absolute value expression to write an inequality for the acceptable values of pH.

68 Convert the absolute value inequality to two inequalities.

69 List two values of pH that will satisfy both inequalities.

A lab technician must classify rock samples according to their ore content. Group I samples have less than 2.5% ore content. Group II samples have from 2.5% to 7.0% ore content. And Group III samples have greater than 7.0% ore content. The table below of the lab's test results shows the ore contents of several rock samples.

Lab Test Results	
Sample ID	**Ore Content**
1	0.3%
2	4.8%
3	2.4%
4	10.0%
5	12.1%
6	5.9%
7	15.7%
8	7.4%
9	2.5%
10	7.0%

70 Write inequalities that describe the percent ore content for each of the three groups.

71 Identify which samples in the table belong to Group I, which belong to Group II, and which belong to Group III.

The specification for the diameter D of a metal rod is

$$2.775 \text{ inches} \pm 0.025 \text{ inches}$$

72 Write a compound inequality that represents the range of values for the rod's diameter D.

73 Use your combined inequality to find the range of dimensions of the radius R of the metal rod.

A technician working for an automobile manufacturer is testing the brake system of new car designs. She uses a formula to estimate braking distance on dry asphalt: $b = 0.039v^2$, where b is the braking distance in feet, and v is the car's speed in miles per hour. The braking test is conducted at three different speeds. A car fails the test if its braking distance exceeds the distance predicted by the formula for any of the three speeds.

74 Write an inequality that describes the maximum braking distance for a car that passes the braking test at a given speed.

75 The braking test is conducted at speeds of 30 miles per hour, 45 miles per hour, and 60 miles per hour. Write a statement that specifies the requirements for passing the entire braking test. (*Hint:* Use the word *and.*)

76 Draw a graph of the inequality in Exercise 74 for speeds up to 60 miles per hour. Show the region that indicates an acceptable braking performance.

Digital computers use logic circuits. Two such circuits are the AND gate and the OR gate. These circuits each have two input lines and one output line. To determine the output, the AND gate "looks" at the two input lines. The output line of an AND gate will be ON only if input line 1 is ON and input line 2 is ON. Similarly, the OR gate determines its output by "looking" at its two input lines. It will turn ON the output line if either line 1 is ON or line 2 is ON. The illustration shows how the AND gate and the OR gate are drawn in circuit diagrams.

A certain circuit controls the two input lines. Input line 1 is ON only when a voltage $V_1 \geq 5.0$ volts. Input line 2 is ON only when the voltage V_2 satisfies the inequality $1.5 \leq V_2 \leq 3.0$ volts. Copy and complete the table below. For each V_1 and V_2, indicate whether line 1 and line 2 will be ON or OFF. Then, for each case, indicate whether the output of the AND gate and the output of the OR gate will be ON or OFF.

	V_1	V_2	Input: Line1	Input: Line2	Output: AND gate	Output: OR gate
77	6.0	6.0				
78	3.0	2.7				
79	4.0	1.2				
80	5.0	2.0				
81	2.0	1.9				
82	7.0	4.0				

A certain assembly requires three identical parts. Each part weighs x grams. After the three parts are connected, five grams are trimmed from the completed assembly. The final assembly must weigh more than 144.1 grams and less than or equal to 161.0 grams.

83 Write an expression for the mass of the final assembly in terms of x, the mass of one part.

84 Write a compound inequality using the result of Exercise 83 and the weight limits of the final assembly.

85 Write your compound inequality as two separate inequalities and solve each one for x.

86 Graph each solution on a number line. Indicate the region that satisfies your combined inequality.

87 Interpret the graph of your compound inequality. What is the range of mass for the parts in the assembly?

You have a box of resistors rated at $\frac{1}{2}$ watt; that is, the resistors can dissipate up to $\frac{1}{2}$ watt of power without being damaged. The resistance values vary from $R = 100$ ohms to $R = 2200$ ohms of resistance. For a resistor of resistance R (in ohms) and carrying a current I (in amperes), the power dissipated is equal to $I^2 R$.

88 Write an inequality stating that the power dissipated by a resistor is less than $\frac{1}{2}$ watt of power.

89 Solve the inequality for the current I for the minimum and maximum resistance values R in the box.

90 Draw a graph of the current I versus the resistance R for the inequality. Shade the region of allowable currents for these resistors.

Susan is an electrician. The wire she is using for a certain construction job comes on 100-foot spools. She uses only $14\frac{1}{2}$-foot lengths and 12-foot lengths of wire on this job.

91 Write an expression for the length of wire used to create a number x of $14\frac{1}{2}$-foot lengths of wire.

92 Write an expression for the length of wire used to create a number y of 12-foot lengths of wire.

93 Write an inequality indicating that the total number of $14\frac{1}{2}$-foot lengths of wire and 12-foot lengths of wire must not exceed the 100 feet of wire on each spool.

Skills

Graph each inequality.

1. $x \geq -1$

2. $-2 < x \leq 2$

Solve each inequality.

3. $2x + 3 \leq 8$

4. $-5x - 4 > 6$

5. $3x < 12$ or $x - 3 > 6$

6. $|x - 4| \leq 2$

Graph each inequality.

7. $x + 2y < 4$

8. $2x - 3y \geq -6$

9. Solve each system. Shade the solution.

$$y - 2x < 5$$
$$2y + x \geq 8$$

Applications

Write an inequality or a system of inequalities to model each situation. Graph the solution.

10. A mechanic has estimated that Carla's repair bill will be less than $250. Carla knows the garage charges $60 for a service call and $45 per hour for the repair work. How many hours did the mechanic estimate the work will take?

11. Clariss needs to bowl an average score of at least 150 to make the finals of the school bowling tournament. She bowled 128 and 160 in her first two games. What score does Clariss need to bowl in her third game to make the finals?

12. Virgil needs to rent a truck and a front-end loader for at least 4 hours to complete a construction project. He can rent a truck for $30 per hour and a front-end loader for $60 per hour. Virgil must keep his expenses under $180. The rental company charges by the whole hour. What combinations of rental times can Virgil use for a truck and a front-end loader in order to satisfy these constraints?

13. The following number line shows the height in centimeters of Janet, Jim, and Martha. Write an inequality to express the relationship of their heights.

14. The specification for a bolt head requires that it be 0.75 ± 0.0015 inches across. Which of the following seven bolts do not meet this specification?

a. 0.7513

b. 0.7401

c. 0.7483

d. 0.7514

e. 0.7519

f. 0.7491

g. 0.7507

15. Write three inequalities that express the following constraints:

- The number of widget x's produced in 30 minutes cannot be less than zero.
- The number of widget y's produced in 30 minutes cannot be less than zero.
- The sum of widget x's and widget y's produced in 30 minutes cannot exceed 45.

Graph the constraints and show the region containing feasible solutions.

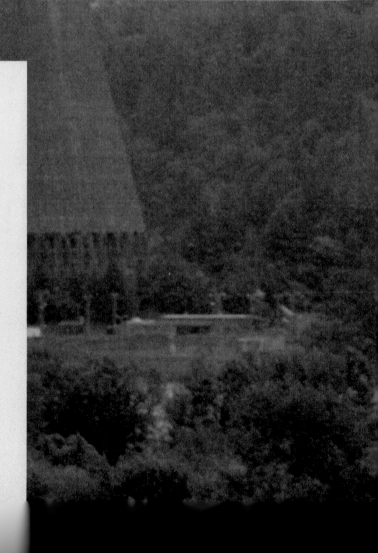

CHAPTER 10

WHY SHOULD I LEARN THIS?

Problem solving is an essential skill in the workplace. The key to solving a problem is dividing it into manageable parts. In this chapter, you will learn to solve complex mathematical expressions by breaking them into simpler parts called factors. If you practice the techniques presented in this chapter, you will develop effective problem-solving strategies.

POLYNOMIALS AND FACTORS

OBJECTIVES

1. Find the product of any two binomials.
2. Factor any trinomial.
3. Solve problems that involve factoring.

The word *factor* is often used in everyday conversations. Consider a few examples. In basketball, you know that height is an important factor in a player's ability to score or block a shot. In a jury trial, a certain piece of evidence may be an important factor in reaching a just verdict. When you plan for a day on the beach, the weather may be a big factor in your decision. The gauges and instruments in a nuclear power plant monitor many important factors related to safety and electrical production. In each of these cases, *factor* means a "part" of something—"part" of an athlete's ability, "part" of the evidence, "part" of a decision-making process or "part" of a power plant's operation.

In mathematics, the word *factor* has a similar meaning. A **factor** is "part" of a mathematical expression. When you factor a mathematical expression, you are writing it in a simpler form. One example of factoring is to use the Distributive Property to rewrite the expression $3x^2 + 5x^2$ as $(3 + 5)x^2$ or $8x^2$. The expression $8x^2$ is simpler than $3x^2 + 5x^2$. The Distributive Property lets you factor out the x^2 term from the original expression.

In this chapter you will see how factoring helps to simplify several different kinds of **polynomials**. In the next chapter you will use the factored form of a polynomial to model many real world situations.

As you watch the video for this chapter, look for examples of how polynomials are factored using geometric models.

Lesson 10.1 Polynomials

A contractor is buying paint to cover the exterior of two cubical storage tanks. Is there an algebraic expression for finding the amount of paint needed to cover both objects?

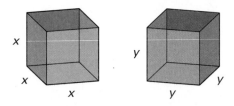

The first storage tank has sides with a length of x feet. The area of one of the faces of this tank is x^2 square feet. Since the tank has six faces, the total exterior area is $6x^2$ square feet. The second storage tank has sides of length y feet and an exterior area of $6y^2$ square feet. The total exterior area of the two tanks is described by the expression

$$6x^2 + 6y^2$$

Monomials, Binomials, and Trinomials

The expression $6x^2 + 6y^2$ contains two monomial terms. A **monomial** is a constant, a variable, or the product of a constant and one or more variables. The sum of one or more monomials is a **polynomial**. For example, $3x^2 + 4x + 2$ is a polynomial in one variable. The monomial terms are $3x^2$, $4x$, and 2.

In this chapter, you will discover patterns that can be represented by three special polynomials.

Name	Number of Terms	Example
Monomial	1	$3x^2$
Binomial	2	$3x^2 + 4x$
Trinomial	3	$3x^2 + 3x + 2$

The expression $3x^2$ is a *mono*mial because it has one term. The expression $3x^2 + 4x$ is a *bi*nomial because it has *two* terms. The expression $3x^2 + 3x + 2$ is a *tri*nomial because it has *three* terms.

Algeblocks can model monomials, binomials, and trinomials. In Chapter 4, you used Algeblocks to model an expression such as $3x + 2$.

To model x^2, a new Algeblock is needed. Use a block that is x units on each side. The area of the block is represented by x^2.

x | $\quad$ x

1 $\qquad$ x

Area $= 1 \cdot x$ $\qquad$ Area $= x \cdot x$
$\quad = x$ $\qquad\qquad = x^2$

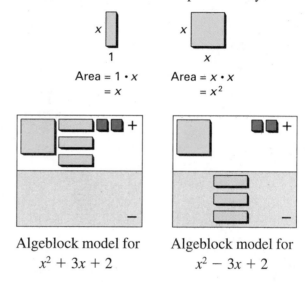

Algeblock model for $\qquad$ Algeblock model for
$x^2 + 3x + 2$ $\qquad\qquad$ $x^2 - 3x + 2$

Adding and Subtracting Polynomials

ACTIVITY | **Adding Trinomials**

1 Use Algeblocks to model $3x^2 + 4x + 1$.

2 Use other Algeblocks to model $2x^2 - 3x + 2$.

3 Combine the Algeblocks.

4 Record your result.

5 What is the sum of $(3x^2 + 4x + 1)$ and $(2x^2 - 3x + 2)$?

6 Complete the equation.

$$(3x^2 + 4x + 1) + (2x^2 - 3x + 2) = ?$$

Recall that monomials that have the same variable factors are called **like terms**. For example, $3x^2$ and $2x^2$ are like terms because each term has x^2 as a common factor. You can add like terms using the Distributive Property.

$$3x^2 + 2x^2 = (3 + 2)x^2 = 5x^2$$

If you add the coefficients of the like terms in the expression

$$(2x^2 + 3x + 4) + (x^2 - 3x + 2)$$

the result is $3x^2 + 6$.

Critical Thinking   Explain how to simplify

$$(2x^2 + 3x + 1) + (x^2 - 3x + 2)$$

EXAMPLE 1 Subtracting Polynomials

Subtract. $(3x^2 - 5) - (x^2 + 2x - 3)$

SOLUTION

You can find the difference by using the basic properties you learned in earlier chapters.

Horizontal Method

$(3x^2 - 5) - (x^2 + 2x - 3) = (3x^2 - 5) + (-1)(x^2 + 2x - 3)$
Definition of subtraction

$= 3x^2 - 5 + (-x^2 - 2x + 3)$
Distributive Property

$= 3x^2 - x^2 - 2x - 5 + 3$
Rearrange terms

$= 2x^2 - 2x - 2$
Add like terms

Vertical Method

$\quad 3x^2 + 0x - 5$ Use zero as coefficient of missing term

$\underline{-(x^2 + 2x - 3)}$ Subtract like terms

$\quad 2x^2 - 2x - 2$

Use any method to simplify $-4x^2 + x - (2x^2 - 3x + 5)$.

EXAMPLE 2 The Perimeter of a Triangle

Find the simplest expression for the perimeter of triangle ABC.

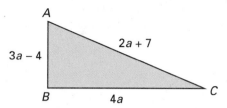

SOLUTION

The perimeter is the total distance around the triangle. To find the perimeter, add the lengths of the three sides; that is, simplify

$$(3a - 4) + (2a + 7) + 4a$$

If you rearrange the terms, the result is $3a + 2a + 4a - 4 + 7$. Now combine like terms. The perimeter is $9a + 3$.

One way to check the solution is to substitute a number for a in both the original expression and in the simplified expression. In this case, substitute 10 for a to see if the results are equal.

$$3a - 4 \quad + \quad 2a + 7 \quad + 4a \quad \stackrel{?}{=} \quad 9a + 3$$

$$[3(10) - 4] + [2(10) + 7] + 4(10) \quad \stackrel{?}{=} \quad 9(10) + 3$$

$$26 \quad + 27 + 40 \quad \stackrel{?}{=} \quad 90 + 3$$

$$93 \quad = \quad 93$$

Since the results are equal, the simplification is correct.

LESSON ASSESSMENT

Think and Discuss

1 Explain what is meant by *like terms* of an expression.

2 Explain the meaning of *monomial*, *binomial*, and *trinomial*. Give an example of each.

3 If *polynomial* means "many names," explain what *polygon* means.

4 How is the Distributive Property used to combine like terms?

Practice and Problem Solving

Classify each of the following expressions as a monomial, binomial, or trinomial.

5. $4x^2 + 3x$ **6.** $15x^4$ **7.** $x - 5$

8. -2 **9.** $7x^2 - 3x + 1$ **10.** $3x$

Use Algeblocks to find each sum.

11. $(3x + 5) + (2x + 3)$

12. $(4x^2 + 3x - 7) + (2x^2 + 5x - 3)$

13. $(3x^2 - 5x + 2) + (4x - 5)$ **14.** $(7x - 5) + (3x - 2)$

Add.

15. $(2x - 7) + (x^2 + 3x - 5)$ **16.** $(3x) + (2x^2 - 3x + 4)$

17. $(3x^2 - 7x + 4) + (x^2 + 7x - 4)$

18. $(4x^2 + 5x - 7) + (3x^2 + 7)$

Subtract.

19. $(2x + 5) - (x + 2)$ **20.** $(5x + 4) - (2x + 7)$

21. $(3x^2 + 8x + 5) - (2x^2 + x + 1)$

22. $(6x^2 + 5x + 7) - (3x - 3)$

23. $(4x^2 - 3x + 4) - (5x + 4)$

24. $(x^2 + 5x - 3) - (3x^2 + 7x + 4)$

25. The sides of a triangle are represented by the expressions $2x^2 + 3$, $4x^2 - 7x$, and $5x + 4$. Write the simplest expression for the perimeter of the triangle.

26. The length of a rectangle is represented by the expression $3x^2 - 2x + 1$. Its width is represented by the expression $x^2 + 2x - 3$. Write the simplest expression for the perimeter of the rectangle.

Mixed Review

27. Solve the distance formula $d = rt$ for the rate r.

28. Find the area of a circular garden with radius 5 meters.

Solve each equation.

29. $2x - 8 = 10$ **30.** $13 = 45 + 8x$ **31.** $3x + 14 = -4x$

32. If the perimeter of a square is 40 cm, what is the length of a diagonal across the square?

33. Barbara bought several CDs for $12 each and paid $5 for a poster. If she spent a total of $41, how many CDs did she buy?

LESSON 10.2 FACTORS AND PRIMES

Factor Pairs

When two or more numbers and/or variables are multiplied together, the numbers or variables are **factors**, and the result is the **product**.

$$a \quad \bullet \quad b \quad = \quad c$$

factor   factor   product

Since $3 \bullet 4 = 12$, 3 and 4 are factors of 12. Although $3 \bullet 4$ and $4 \bullet 3$ are two different multiplication facts, they are considered the same **factor pair**. This text lists factor pairs with the smaller number written first. Thus, you can write the factor pairs of 12 as

$$1 \bullet 12, \ 2 \bullet 6, \text{ and } 3 \bullet 4$$

The area of a rectangle is the product of two factors (the length and the width). Thus, you can use an Algeblock model or an area diagram to represent a factor pair.

3 blocks

→| 4 blocks |←

3 rows of 4 blocks each = 12 blocks
$(3 \times 4) = 12$

Algeblock Model

3 cm

4 cm

$A = 12 \text{ cm}^2$
$(3 \times 4) = 12$

Area Diagram

ACTIVITY   **Building Rectangles with Factor Pairs**

1   Use Algeblocks or area diagrams to model the factor pairs for each whole number between 1 and 13. Record the results in a table like the one shown on the opposite page.

Number	Factor Pairs	Number of Factor Pairs
2	?	?
3	?	?
4	?	?

and so on until you reach the following:

12	1 • 12; 2 • 6; 3 • 4	3

2 Which numbers have exactly one factor pair? These numbers are **prime numbers**.

> **Prime Numbers**
> If a whole number, p, greater than 1 has exactly two unique whole-number factors, 1 and the number itself, then p is a prime number.

3 Which numbers have a factor pair in which both numbers are the same? These numbers are called **perfect square numbers**.

4 Which numbers have two or more factor pairs? These numbers are called **composite numbers**.

5 Continue the chart to find all of the prime numbers less than 30.

You can use factor pairs to write the factors of a number. For example, the factor pairs of 12 are

$$1 • 12, 2 • 6, \text{ and } 3 • 4$$

Thus, the factors of 12 are

$$1, 2, 3, 4, 6, \text{ and } 12$$

Ongoing Assessment

What are the factor pairs for 18? List the factors of 18.

Prime Factorization

A number is **factored** when it is written as a product of two of its factors. A number is *completely factored* when it is written as a product of prime factors.

EXAMPLE 1 Listing Prime Factors

List the prime factors of 24.

SOLUTION

Start with 24 and use a factor tree to write the factor pairs. Begin by writing down any factor pair of 24.

24

4 • 6

Continue the tree.

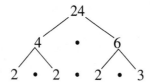

The prime factors of 24 are 2 and 3.

When you factor 24 into 2 • 2 • 2 • 3, you are writing the **prime factorization** of 24. You can also write the prime factorization of 24 using exponents. Thus,

$$2 \cdot 2 \cdot 2 \cdot 3 = 2^3 \cdot 3$$

EXAMPLE 2 Prime Factorization

Write the prime factorization of 36.

SOLUTION

Start with 36 and build a factor tree.

Thus, $36 = 2 \cdot 2 \cdot 3 \cdot 3$

$\quad\quad\ = 2^2 \cdot 3^2$

Since 2 and 3 are prime numbers, $2^2 \cdot 3^2$ is the prime factorization of 36.

Ongoing Assessment

Find the prime factorization of 36 by starting with a factor pair that is different than the one used in the example.

Greatest Common Factors

The numbers 24 and 36 have several **common factors**.

Factors of 24	**1, 2, 3, 4, 6, 12,** 24
Factors of 36	**1, 2, 3, 4, 6,** 9, **12,** 18, 36

The common factors of 24 and 36 are **1, 2, 3, 4, 6,** and **12**. The **greatest common factor** of 24 and 36 is 12.

> ### Greatest Common Factor (GCF)
> The greatest common factor of two or more integers is the largest factor that all the integers share.

You can use prime factorization to find the greatest common factor of a set of numbers. The GCF is the product of the common prime factors of the numbers in the set.

$$24 = 2 \cdot \mathbf{2 \cdot 2 \cdot 3}$$

$$36 = \quad \mathbf{2 \cdot 2 \cdot 3} \cdot 3$$

Thus, the GCF is $\mathbf{2 \cdot 2 \cdot 3}$, or 12.

In the next section, the idea of GCF will be extended to monomials and polynomials.

CULTURAL CONNECTION

Many people have tried to find a formula for generating the set of all prime numbers. One such formula, $n^2 - n + 41$, will generate primes by substituting whole numbers less than 41 for n. Can you explain why the formula breaks down at $n = 41$?

An early Greek mathematician, Eratosthenes, used a method called a sieve to find prime numbers.

List the numbers 1–10 in the first row of a table. Then, continue listing the numbers 11–100 in rows of 10. Start at 2 and cross out every second number except 2. Start at 3 and cross out every third number except 3. Continue through all the rows of the table until no more numbers can be crossed out. Describe the set of numbers remaining.

Prime numbers such as three and five are called **twin primes** because their difference is two. A famous unproved theorem states that there is an infinite number of twin primes. Use the Sieve of Eratosthenes to list the twin primes less than 100.

LESSON ASSESSMENT

Think and Discuss

1 What are factor pairs?

2 How are factor pairs used in prime factorization?

3 What is the difference between a prime number and a composite number?

4 What is the difference between the prime factors of a number and the prime factorization of a number?

5 What is the greatest common factor of a set of numbers?

6 Can a set of numbers have more than one common factor?

7 Give an example.

What are the factor pairs for each number?

7. 45   **8.** 17   **9.** 32

10. 27   **11.** 26   **12.** 33

Write the prime factorization of each number.

13. 42   **14.** 35   **15.** 18

16. 84   **17.** 28   **18.** 51

Use prime factorization to find the greatest common factor of each pair of numbers:

19. 6 and 12   **20.** 5 and 7   **21.** 36 and 60

22. 50 and 52   **23.** 39 and 52   **24.** 72 and 90

25. Find the greatest common factor of 18, 36, and 60.

Mixed Review

Solve for x and y.

26. $3x - y = 7$ **27.** $3x - 4y = 9$ **28.** $2x - 3y = 12$
      $2x + 3y = 12$ $2x - 3y = 7$ $5x + 2y + 8 = 0$

29. An iron beam 68 inches long is divided into two parts. One piece is three inches longer than the other. What are the two lengths?

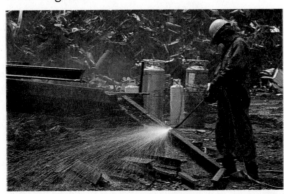

30. Find the angles of a triangle if one angle is three times a second angle and six times the third angle.

LESSON 10.3 MONOMIAL FACTORS

The Distributive Property

An area diagram is a geometric model of the Distributive Property.

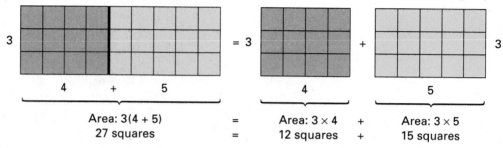

Area: 3(4 + 5) = Area: 3 × 4 + Area: 3 × 5
27 squares = 12 squares + 15 squares

In general, you can multiply a monomial a and a binomial $(b + c)$ using the Distributive Property. The model looks like this.

Area: $a(b + c)$ = Area: ab + Area: ac

Likewise, you can use the Distributive Property with expressions that contain exponents.

$$3x^2(x^3 + 4) = 3x^2(x^3) + 3x^2(4)$$

To simplify the expression $3x^2(x^3)$, you can use the properties of exponents you learned in Chapter 2.

EXAMPLE 1 Adding Exponents

Simplify $x^2 \cdot x^3$

SOLUTION

$$x^2 \cdot x^3 = (x \cdot x) \cdot (x \cdot x \cdot x)$$

$$= x \cdot x \cdot x \cdot x \cdot x$$

$$= x^5$$

Notice that the exponent for the product is the sum of the exponents of the factors; that is, $x^2 \cdot x^3 = x^{2+3} = x^5$. This property is the Product of Powers Property.

Product of Powers Property

For all numbers x and all integers m and n,

$$x^m \cdot x^n = x^{m+n}$$

Now the simplification can be completed.

$$3x^2(x^3 + 4) = 3x^2 (x^3) + 3x^2 (4) = 3x^5 + 12x^2 \quad \blacksquare$$

Ongoing Assessment

The electrical power of a solar car is found using the equation $P_E = I(V - IR)$. Use the Distributive Property to multiply the terms on the right side of the equality. $\quad \blacksquare$

Some polynomials have a common monomial factor. For example, the binomial $20H^2 + 50H$ estimates part of a company's cost of renting advertising space on a billboard. Notice that each term in the binomial contains the common factor $10H$.

Use the Distributive Property to factor out the common monomial.

$$20H^2 + 50H = (10H \cdot 2H) + (10H \cdot 5)$$
$$= 10H(2H + 5)$$

To *factor* $20H^2 + 50H$ means to write it as $10H(2H + 5)$. When you factor a polynomial, be sure to identify *all* the common factors.

EXAMPLE 2 Removing Common Monomial Factors

The U.S. Forest Service uses aerial photography in a survey to estimate the diameter of trees. If H is the height of a tree and V is the visible crown diameter from the photograph, then the tree's diameter is given by the polynomial $2HV - 2HV^2 - 6HV^3$. Factor this polynomial.

SOLUTION

To identify the common monomial factors, factor each term of the polynomial.

$$(2 \cdot H \cdot V) - (2 \cdot H \cdot V \cdot V) - (2 \cdot 3 \cdot H \cdot V \cdot V \cdot V)$$

The greatest common monomial factor is $2HV$. Factor out $2HV$ and write

$$2HV - 2HV^2 - 6HV^3 = 2HV(1 - V - 3V^2) \quad \blacksquare$$

Factor the expression $4x^3 + 8x^2 + 12x$.

Dividing by a Monomial

The Distributive Property is true for all numbers, including the rational numbers. Thus, you can use the Distributive Property to simplify an expression such as $\frac{1}{x^2}(x^5 + x^3)$ where $x \neq 0$.

$$\frac{1}{x^2}(x^5 + x^3) = \frac{1}{x^2}(x^5) + \frac{1}{x^2}(x^3)$$

$$= \frac{x^5}{x^2} + \frac{x^3}{x^2}$$

To simplify the right side of the equation, you can use another property of exponents from Chapter 2.

EXAMPLE 3 Subtracting Exponents

Simplify $\frac{x^5}{x^2}$.

SOLUTION

$$\frac{x^5}{x^2} = \frac{x \cdot x \cdot x \cdot \cancel{x} \cdot \cancel{x}}{\cancel{x} \cdot \cancel{x}}$$

$$= x^3$$

What do you notice about the difference between the exponents in the original expression and the exponent in the answer?

Quotient of Powers Property
For all non-zero numbers x and integers m and n,

$$\frac{x^m}{x^n} = x^{m-n}$$

Now you can simplify the right side of the equation above.

$$\frac{x^5}{x^2} + \frac{x^3}{x^2} = x^{5-2} + x^{3-2} = x^3 + x$$

Simplifying $\frac{1}{x^2}(x^5 + x^3)$ where $x \neq 0$, is the same as dividing the expression $(x^5 + x^3)$ by the monomial x^2. Remember, when you divide an expression by a monomial, divide each of the terms by the monomial.

Ongoing Assessment

An electronics technician analyzes a simple circuit and verifies that the equation below models the voltage V and current I in the circuit, where the resistance R and k are constant.

$$VI = I^2R_1 + I^2R_2$$

a. Factor the expression on the right side of the equation.

b. Solve the equation for V.

LESSON ASSESSMENT

Think and Discuss

1 How is the Distributive Property used to multiply a monomial by a polynomial?

2 How is the Distributive Property used in factoring?

Practice and Problem Solving

Multiply.

3. $a(b + 3)$ **4.** $x(y - 5)$ **5.** $7x(x + 5y)$

6. $3x^2(4x^2 + 7)$ **7.** $2x(4x^3 + 7x - 5)$ **8.** $x^4(x^2 - 2x + y)$

Divide.

9. $4x^2 + 8y^2$ by 4 **10.** $2x^3 + 4x$ by $2x$

11. $3x^4y^2 - 6x^2y^4$ by $3xy$

12. An automated assembly line is programmed to produce a variety of sizes of rectangular toolboxes. The volume of a toolbox produced on this line is given by $6ab^2$. If the width is $3b$ and the depth is a, what is the height of this toolbox?

The volume of a cylinder of radius r and height h is
$V = \pi r^2 h$.

13. Write an equation in simplified form for the volume of air in a scuba diver's cylindrical tank if the height of the tank is three times its radius.

14. Rewrite this equation for the diameter (d) of the tank and simplify.

Factor out the greatest monomial factor.

15. $3xy - 5y^2$ **16.** $-3ax + cx - x^2$ **17.** $cx^2 + 2cx^3$

Mixed Review

A jar contains 7 pennies, 8 nickels, and 5 dimes.

18. If one coin is taken at random from the jar, what is the probability it will be a penny?

19. If one coin is taken at random from the jar, what is the probability it will be a nickel or a dime?

20. If two coins are taken at random from the jar, what is the probability that both coins are dimes?

Describe the set of whole numbers that satisfies each of the following inequalities:

21. $3 + n < 10$

22. $n + 2 > 8$

23. $3 + n < 9$ and $2 + n < 6$

Write and solve an equation for each situation.

24. The sum of three consecutive integers is 39. What is the middle integer?

25. A bank charges $3 for every $500 worth of certified checks it issues. If the charge for Mary's certified check is $12, what is the value of the check issued by the bank?

LESSON 10.4 ADDITIONAL PROPERTIES OF EXPONENTS

Zero and Negative Exponents

The following Example will help you recall two of the properties of exponents used in Chapter 2.

EXAMPLE 1 Zero Exponents

How can you convince someone that $x^0 = 1$?

SOLUTION

Show that the expression $x^6 \div x^6$ can be written as either 1 or x^0.

a. Factor the numerator and denominator.

$$\frac{x^6}{x^6} = \frac{\cancel{x} \cdot \cancel{x} \cdot \cancel{x} \cdot \cancel{x} \cdot \cancel{x} \cdot \cancel{x}}{\cancel{x} \cdot \cancel{x} \cdot \cancel{x} \cdot \cancel{x} \cdot \cancel{x} \cdot \cancel{x}} = \frac{1}{1} = 1$$

b. Use the Quotient of Powers Property.

$$\frac{x^6}{x^6} = x^{6-6} = x^0$$

Thus, $x^0 = 1$.

This example leads to the following property.

> **Zero Property of Exponents**
> For all numbers x (except zero), $x^0 = 1$

EXAMPLE 2 Negative Exponents

How can you convince someone that $\dfrac{1}{x^2} = x^{-2}$?

SOLUTION

Start with $\dfrac{x^3}{x^5}$.

a. Factor the numerator and denominator.

$$\frac{x^3}{x^5} = \frac{\cancel{x} \cdot \cancel{x} \cdot \cancel{x}}{\cancel{x} \cdot \cancel{x} \cdot \cancel{x} \cdot x \cdot x} = \frac{1}{x \cdot x} = \frac{1}{x^2}$$

b. Use the Quotient of Powers Property.

$$\frac{x^3}{x^5} = x^{3-5} = x^{-2}$$

Thus, $x^{-2} = \dfrac{1}{x^2}$.

This example leads to the following property.

> ### Property of Negative Exponents
> For all numbers x (except zero) and integers n,
>
> $$x^{-n} = \frac{1}{x^n}$$

Critical Thinking   Explain why x^{-1} and x are reciprocals.

Powers of Products and Quotients

In some problems, it is necessary to raise a product or a fraction to a power before you can find a solution.

EXAMPLE 3   Raising a Monomial to a Power

Simplify $(5x)^3$.

SOLUTION

$$(5x)^3 = 5x \cdot 5x \cdot 5x$$
$$= (5 \cdot 5 \cdot 5)(x \cdot x \cdot x)$$
$$= 5^3 \cdot x^3$$
$$= 125x^3$$

Notice that when the product is raised to a power, each factor is raised to that same power.

> ### Power of a Product
> For all numbers x and y, and integers n,
>
> $$(xy)^n = x^n y^n$$

EXAMPLE 4 Power Dissipation

An electric component dissipates power (in watts) at a rate given by the following formula:

$$P = I^2R$$

In this formula, I is the current passing through the component (in amperes) and R is the resistance of the component (in ohms). An electronics technician needs to predict the heat that will be generated by a resistor on a control board. He measures a current of 0.1 amperes passing through the resistor. The resistor has a resistance of 300 ohms. The power dissipated as heat is

$$P = I^2R$$
$$= (0.10^2)(300)$$
$$= 3 \text{ watts}$$

What power will the same resistor dissipate if the current doubles?

SOLUTION

Let I and R be the same numerical values as before. This time, the current is $2I$. Substitute $2I$ for I and simplify.

$$P = (2I)^2R$$
$$= (2)^2(I)^2R$$
$$= 4I^2R$$

If the current doubles, the power quadruples.

What happens when a fraction is raised to a power?

EXAMPLE 5 The Power of a Quotient

Simplify $\left(\dfrac{x}{3}\right)^3$.

SOLUTION

$$\left(\frac{x}{3}\right)^3 = \frac{x}{3} \cdot \frac{x}{3} \cdot \frac{x}{3}$$
$$= \frac{x \cdot x \cdot x}{3 \cdot 3 \cdot 3}$$
$$= \frac{x^3}{27}$$

When a fraction is raised to a power, the numerator and the denominator are each raised to the same power.

Power of a Quotient

For all numbers x and y (except zero) and integers n,

$$\left(\frac{x}{y}\right)^n = \frac{x^n}{y^n}$$

Ongoing Assessment

The resistor in Example 4 dissipated 3 watts of power with a current of 0.1 amperes. Use the Power of a Quotient Property to find the power dissipated when the current is halved.

Power of a Power

EXAMPLE 6 Raising a Power to a Power

Simplify $(x^3)^2$.

SOLUTION

$$(x^3)^2 = x^3 \bullet x^3$$
$$= x^{3+3}$$
$$= x^6$$

Notice that the product of the original two exponents $2 \bullet 3$ also equals 6. Thus,

$$(x^2)^3 = x^{2 \bullet 3} = x^6$$

When a power is raised to a power, the two powers are multiplied.

Power of a Power

For all numbers x and integers n and m,

$$(x^n)^m = x^{nm}$$

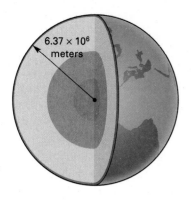

6.37 × 10⁶
meters

EXAMPLE 7 The Volume of the Earth

The radius of the Earth is 6.37×10^6 meters. What is its volume? Write the answer in scientific notation.

SOLUTION

The Earth is approximately a sphere. The volume of a sphere is found by using a formula.

$$V = \frac{4}{3}\pi r^3$$
$$= \frac{4}{3}\pi(6.37 \times 10^6)^3$$
$$= \frac{4}{3}\pi(6.37)^3(10^6)^3$$
$$= \frac{4}{3}\pi(6.37)^3 \times 10^{18}$$
$$\approx 1083 \times 10^{18}$$
$$\approx 1.08 \times 10^{21}$$

The volume of the Earth is approximately 1.08×10^{21} cubic meters.

WORKPLACE COMMUNICATION

What electrical current should the maintenance technician expect from the radiometer when measuring the furnace temperature at $\frac{1}{2}T_{ave}$ and $\frac{3}{2}T_{ave}$?

INSTRUMENTATION MAINTENANCE REQUEST FORM

SUBMITTED BY: Melanie Montoya

EXTENSION: 6-0318

OFFICE: Semiconductor Dev Team

EMAIL ADDRESS: melmont@mwi.org

DATE/TIME SUBMITTED: Oct 22/9:30 AM

DATE/TIME NEEDED: Oct 23/5:00 PM

COMPLETE DESCRIPTION OF INSTRUMENTATION MAINTENANCE REQUIRED:

We have built an instrument for measuring the temperature of the furnace in the semiconductor production area. This instrument is a *Stefan-Boltzmann* radiometer. The radiometer detects heat energy radiated from the furnace and converts it into an electrical current. This conversion is described by the formula

$I = CT^4$

where I is the instrument current, T is the furnace temperature, and C is a constant.

In a recent test, we found that the radiometer produces 1 milliampere of current when the furnace is at the average temperature (T_{ave}).

Please calibrate the radiometer so that it will accurately read temperatures from a low of $\frac{1}{2}T_{ave}$ to a high of $\frac{3}{2}T_{ave}$.

TO BE COMPLETED BY MAINTENANCE TECHNICIAN

DATE/TIME RECEIVED: Oct 22/11:00 AM

SIGNATURE:

DATE/TIME COMPLETED:

LESSON ASSESSMENT

Think and Discuss

1 Explain why $(2x)^3$ is not the same as $2x^3$.

2 Suppose the side of a square is doubled. Explain why the area of the square is four times as large as the original square.

3 Suppose the edge of a cube is doubled. What is the effect on the volume of the cube? Explain your answer.

4 Explain why $\left(\dfrac{a}{b}\right)^{-n}$ is the same as $\left(\dfrac{b}{a}\right)^{n}$.

Practice and Problem Solving

Simplify each expression.

5. $a^2\,(a^5)$

6. $(-3x)^3$

7. $2x^3\,(5x^4)$

8. $(3x^2)^4$

9. $\left(\dfrac{2x}{3}\right)^3$

10. $(5y^{-2})^{-3}$

11. $2x^3\,(3x^2 + 4)$

12. $5xy\,(3x^2 + 2xy - 4y^3)$

13. $\dfrac{16s^3t^2}{(2st)^2}$

Mixed Review

Solve each problem. Write an equation to model each situation.

14. A circular pipe has a radius of 3 inches. The pipe is placed inside a square box that has an inside dimension of 6 inches. What is the cross-sectional area outside the pipe but inside the box?

15. If 48 pounds of steel cost D dollars, write a formula that determines the cost of x pounds of steel.

16. A group of 36 people is divided into two groups. The first group is four-fifths the size of the second group. How many people are in each group?

17. The perimeter of a rectangle is 36 inches. The length is 4 inches longer than the width. Find the area of the rectangle.

18. The sum of two angles is 180 degrees. The ratio of the angle measures are two to one. What are the measures of the angles?

19. Solve the system $x > 2$ and $x + y \leq 5$ by graphing.

LESSON 10.5 ADDING RATIONAL EXPRESSIONS

Tammy can fill a tank in 10 hours by opening Pipe A. In one hour, $\frac{1}{10}$ of the tank is filled. In 3 hours, the tank is $3 \cdot \frac{1}{10}$ or $\frac{3}{10}$ full, and in x hours, $x \cdot \frac{1}{10}$ or $\frac{x}{10}$ of the tank is full. In 10 hours, $\left(10 \cdot \frac{1}{10} = 1\right)$ the tank is completely full. Pipe B can fill the tank by itself in 6 hours. How can Tammy find the time it will take to fill the tank if she opens both pipes? Let x represent the time it takes both pipes to fill the tank. In x hours, Pipe A will fill $\frac{x}{10}$ of the tank, and Pipe B will fill $\frac{x}{6}$ of the tank. To find the time to fill the tank, add the contributions of both pipes.

$$\frac{x}{10} + \frac{x}{6} = 1$$

Adding rational expressions is like adding rational numbers. If necessary, first rewrite the terms with a common denominator. Then use the following rule:

Adding Rational Expressions

If $\frac{a}{b}$ and $\frac{c}{b}$ are rational expressions ($b \neq 0$), then

$$\frac{a}{b} + \frac{c}{b} = \frac{a + c}{b}$$

To solve Tammy's problem, start by writing the rational expressions with common denominators. Thus,

$$\frac{3}{3} \cdot \frac{x}{10} + \frac{5}{5} \cdot \frac{x}{6} = 1$$

Now add the rational expressions and solve for x.

$$\frac{3x}{30} + \frac{5x}{30} = 1$$
$$\frac{8x}{30} = 1$$
$$8x = 30$$
$$x = \frac{30}{8} \quad \text{or} \quad 3\frac{3}{4}$$

It will take $3\frac{3}{4}$ hours to fill the tank with both pipes open.

EXAMPLE 1 Completing a Job Alone

Ron and Al are preparing an estimate to repair a car. They estimate it will take 6 hours to repair the car if they work on it together. Ron can complete the repair by himself in 10 hours. How long will it take Al to do the job alone?

SOLUTION

Let x represent the amount of time it takes Al to do the job alone.

Then $\frac{1}{x}$ represents the part of the job that Al can complete in 1 hour. Remember that the denominator cannot be zero. In 6 hours, Al can complete $\frac{1}{x} \cdot 6$, or $\frac{6}{x}$ of the total job.

Ron can complete $\frac{1}{10}$ of the job in 1 hour. In 6 hours, he can complete $\frac{6}{10}$, or $\frac{3}{5}$ of the job.

The total job, represented by 1, is equal to the sum of Al's part and Ron's part.

The equation that models this situation is
$$\frac{6}{x} + \frac{3}{5} = 1$$

Write the rational expressions with common denominators. Then add the expressions and solve for x.

$$\frac{5}{5} \cdot \frac{6}{x} + \frac{x}{x} \cdot \frac{3}{5} = 1$$

$$\frac{30}{5x} + \frac{3x}{5x} = 1$$

$$\frac{30 + 3x}{5x} = 1$$

$$30 + 3x = 5x$$

$$30 = 2x$$

$$x = 15$$

It will take Al 15 hours to repair the car by himself. Check to see if this is the correct answer.

Ongoing Assessment

It takes two people 4 hours to clean an apartment if they work together. One person can clean the apartment in 6 hours. How long will it take the other person to clean the apartment working alone?

EXAMPLE 2 Adding Rational Expressions

Simplify $\dfrac{3}{x} + \dfrac{4}{x^2}, x \neq 0$.

SOLUTION

First, rewrite the expressions with a common denominator. Then add the expressions.

$$\frac{x}{x} \cdot \frac{3}{x} + \frac{4}{x^2} = \frac{3x}{x^2} + \frac{4}{x^2}$$

$$= \frac{3x + 4}{x^2}$$

The final expression $\dfrac{3x + 4}{x^2}$ carries the restriction that $x \neq 0$.
When you solve a rational equation, this restriction is very important.

Ongoing Assessment

Simplify $\dfrac{a}{x^2} + \dfrac{1}{x^3}, x \neq 0$.

You can also subtract rational expressions. Use the addition rule with the addition sign changed to a subtraction sign.

EXAMPLE 3 Subtracting Rational Expressions

a. Simplify $\dfrac{3}{2x} - \dfrac{1}{5x}, x \neq 0$.

b. Solve $\dfrac{3x}{x - 2} - \dfrac{5}{2 - x} = 1, x \neq 2$.

SOLUTION

a. The least common denominator is $10x$. Thus,

$$\frac{3}{2x} - \frac{1}{5x} = \frac{5}{5} \cdot \frac{3}{2x} - \frac{2}{2} \cdot \frac{1}{5x}$$

$$= \frac{15}{10x} - \frac{2}{10x}$$

$$= \frac{13}{10x}, x \neq 0$$

b. The denominators are opposites. Multiply the first rational expression by $\frac{-1}{-1}$ so that the rational expressions will have common denominators. The result is

$$\frac{-1}{-1} \cdot \frac{3x}{x-2} = \frac{-3x}{-(x-2)} = \frac{-3x}{2-x}$$

Now subtract and solve the equation for x.

$$\frac{-3x}{2-x} - \frac{5}{2-x} = 1$$

$$\frac{-3x-5}{2-x} = 1$$

$$-3x - 5 = 2 - x$$

$$-2x = 7$$

$$x = -3.5$$

Check to see if -3.5 is a solution to the original equation.

LESSON ASSESSMENT

Think and Discuss

1 What is a rational expression?

2 How do you add rational expressions?

3 How do you subtract rational expressions?

4 Why is it necessary to use restrictions when you add and subtract rational expressions?

Practice and Problem Solving

Add or subtract.

5. $\dfrac{x}{2} + \dfrac{3x}{4}$

6. $\dfrac{x+1}{3} - \dfrac{2x-1}{4}$

7. $\dfrac{3}{2x} - \dfrac{4}{5}$

8. $\dfrac{x+5}{2x} + \dfrac{3}{x}$

9. $\dfrac{x}{2} + \dfrac{1}{x}$

10. $\dfrac{x-1}{4x} + \dfrac{3-x}{6x}$

11. $\dfrac{x+3}{x^2} - \dfrac{2x-5}{x^2}$

12. $\dfrac{7}{x} + \dfrac{3x+1}{x^2}$

13. $\dfrac{x^2+2}{x^2} - \dfrac{x-1}{-2x}$

14. A farmer owns a tractor that can cultivate a field in 8 hours. The farmer is trying to decide whether to buy another tractor that can cultivate the same field in 6 hours. How long would it take to cultivate the field using both tractors?

15. Miquel and Rosa are painting contractors. They spent 5 hours painting a room together. Last year Miquel painted the same room in 10 hours. How long would it take Rosa to paint the room by herself?

16. A company is preparing a mass mailing to advertise its new stereo system. One employee can complete the address labels for the mailing in 4 hours. Another employee can do the job in 3 hours. If both employees work together, how long will it take to prepare the address labels?

17. Two electricians wired a new house in 30 hours. On a similar house, one of the electricians spent 50 hours to install all the wiring. How long would it take the other electrician to wire the house if he worked alone?

Mixed Review

Solve each problem.
18. Althea paid $250 for her utility bills in May and June. In June, the bill was $25 plus twice her bill for May. How much was Althea's utility bill for each month?

19. There are 91 students in three classes. One class has 9 more students than the first class. The other class has 5 fewer students than the first class. How many students are there in each class?

20. After Ronald deposited $125.50 into his savings account, he still had less than $200 in his account. How much did Ronald have in his account before he made the deposit?

21. The cost of three shirts and two pairs of shorts is $149. The cost of two shirts and three pairs of shorts is $136. Find the cost of the shirts and the shorts.

22. A 25% soap solution is mixed with a 40% soap solution to obtain 16 gallons of a 30% solution. How many gallons of each soap solution are needed?

A small patio is 4 feet by 6 feet. You want to expand the patio by the same length on each side. What formula expresses the new area?

4 feet

6 feet

Examine the diagram of the patio. Let x represent the amount added to each side of the patio. Then one side will be represented by $x + 4$ and the other side by $x + 6$. The expression for the new area of the patio will be the product of the binomials.

$$(x + 4)(x + 6)$$

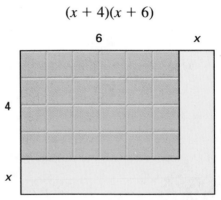

6   x

4

x

Notice you can divide the new area into four regions with the following areas:

$$x^2 \quad 6x \quad 4x \quad 24$$

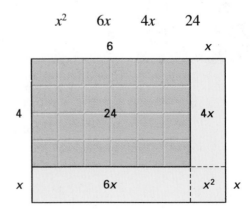

6   x

4   24   4x

x   6x x^2 x

The total area is the sum of the areas:

$$x^2 + 6x + 4x + 24 \quad \text{or} \quad x^2 + 10x + 24$$

Suppose you need to expand the patio by 3 feet on each side. Substitute 3 for x to check if $(x + 4)(x + 6) = x^2 + 10x + 24$.

$$(3 + 4)(3 + 6) = 3^2 + 10(3) + 24$$

$$(7)(9) = 9 + 30 + 24$$

$$63 = 63$$

EXAMPLE 1 Multiplying Two Binomials

Multiply $(x + 2)(x + 1)$.

SOLUTION

The product $(x + 2)(x + 1)$ can represent a rectangle with length $(x + 2)$ and width $(x + 1)$. The area of this rectangle is

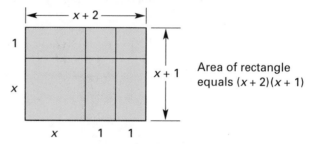

Area of rectangle equals $(x + 2)(x + 1)$

If you label the area for each of the small rectangles, you can determine the total area of the overall rectangle. This area is the same as the product of $(x + 2)(x + 1)$,

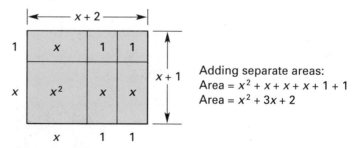

Adding separate areas:
Area $= x^2 + x + x + x + 1 + 1$
Area $= x^2 + 3x + 2$

You have just used a geometric representation to multiply two binomials, $x + 2$ and $x + 1$. The result is the trinomial $x^2 + 3x + 2$.

Thus, $(x + 2)(x + 1) = x^2 + 3x + 2$.

The geometric model can be used to derive an algebraic pattern for multiplying two binomials. Consider the product $(x + 3)(2x + 1)$.

Let $(x + 3)$ be the length of a rectangle and $(2x + 1)$ be the width.

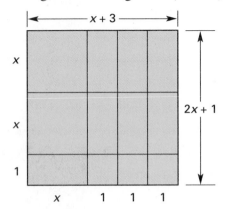

The development of the algebraic product $(x + 3)(2x + 1)$, term by term, is shown in a series of four drawings. The geometric equivalent of each term-by-term product is highlighted by the shaded portions in the rectangle.

Step 1: $(\overparen{x + 3})(\overparen{2x} + 1)$
$(x)(2x) = 2x^2$

Step 2: $(\overparen{x} + 3)(2x + \overparen{1})$
$(x)(1) = x$

Step 3: $(x + \overparen{3})(\overparen{2x} + 1)$
$(3)(2x) = 6x$

Step 4: $(x + \overparen{3})(2x + \overparen{1})$
$(3)(1) = 3$

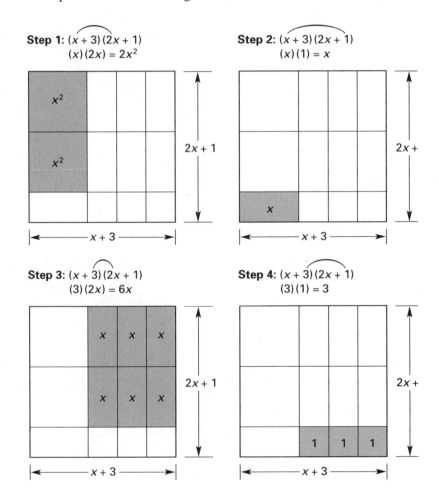

Add together all the highlighted areas in the four rectangles to find the total area of the rectangle of size $(x + 3)$ by $(2x + 1)$.

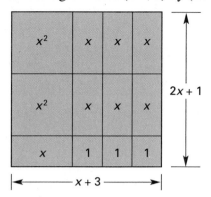

$$(x + 3)(2x + 1) = x^2 + x^2 + x + x + x + x + x + x + x + 1 + 1 + 1$$

$$= 2x^2 + 7x + 3$$

When the four steps in the multiplication process are done in sequence, the result is the following:

$$(x + 3)(2x + 1) = 2x^2 + x + 6x + 3$$

$$(x + 3)(2x + 1) = 2x^2 + 7x + 3$$

A simple memory device will help you keep track of which terms are being multiplied. This device is known as the FOIL technique.

To find the product of $(x + 3)(2x + 1)$, proceed as follows:

First terms are multiplied $\quad (\boldsymbol{x} + 3)(\boldsymbol{2x} + 1) \quad x \bullet 2x = 2x^2$

Outer terms are multiplied $\quad (\boldsymbol{x} + 3)(2x + \boldsymbol{1}) \quad x \bullet 1 = x$

Inner terms are multiplied $\quad (x + \boldsymbol{3})(\boldsymbol{2x} + 1) \quad 3 \bullet 2x = 6x$

Last terms are multiplied $\quad (x + \boldsymbol{3})(2x + \boldsymbol{1}) \quad 3 \bullet 1 = 3$

Add and then simplify by combining similar terms. The result is

$$(x + 3)(2x + 1) = 2x^2 + x + 6x + 3$$

$$= 2x^2 + 7x + 3$$

The FOIL method is a shortcut for using the Distributive Property. It helps you multiply two binomials in an orderly manner. Practice will help make the process automatic.

When one or both binomials in a product contains a negative term, the geometric model becomes complicated to use. For example, if you are asked to find the product $(x - 2)(x + 1)$, it is hard to imagine how to represent -2.

However, the FOIL method readily handles the negative terms.

$$\overset{\text{F}}{}\quad\overset{\text{O}}{}\quad\overset{\text{I}}{}\quad\overset{\text{L}}{}$$
$$(x - 2)(x + 1) = (x)(x) + (x)(1) + (-2)(x) + (-2)(1) = x^2 - x - 2$$

Algeblocks are another way to use areas to model the multiplication of binomials. Algeblocks solve the problem of negative areas by using a quadrant system to identify areas as either positive or negative.

EXAMPLE 2 Using Algeblocks to Multiply Binomials

Multiply $(x - 2)(x + 1)$ using Algeblocks as a model.

SOLUTION

1. Select the blocks that model $x - 2$ and $x + 1$.

2. Arrange these blocks along the axes on the Algeblock mat as shown in the diagram. Note that the 2 unit blocks that model -2 are placed along the negative portion of the horizontal axis.

3. Now use additional Algeblocks to form a rectangle using the blocks on the axes to model the length and width of the rectangle. Note that the *areas* in the second quadrant are labeled negative. List the six *areas* that make up the rectangle you formed.

4. Add the *areas* of the rectangle formed by the Algeblocks. Do not include in the sum the *areas* of the Algeblocks on the axes. What is the sum of the *areas*? What is the product of $(x - 2)$ and $(x + 1)$?

Critical Thinking Explain how the Distributive Property is used to find the following product:

$$(x + 3)(2x + 1) = (x + 3)(2x) + (x + 3)(1)$$

$$= (x)(2x) + (3)(2x) + (x)(1) + (3)(1)$$

$$= 2x^2 + 6x + x + 3$$

$$= 2x^2 + (6 + 1)x + 3$$

$$= 2x^2 + 7x + 3$$

LESSON ASSESSMENT

Think and Discuss

1 How does the geometric model illustrate the product of two binomials? Show an example.

2 How do you use the FOIL method to find the product of two binomials? Show an example.

3 Try doing the multiplication by placing one binomial under the other and multiplying as if they were whole numbers. How is this method similar to numerical multiplication?

4 Use the FOIL method to find the product $(10 + 4)(10 + 6)$. Explain how the FOIL method can be used to find the product of two numbers such as $(24)(35)$.

Find the products using a geometric model.

5. $(x + 4)(x + 3)$ **6.** $(x + 5)(x + 1)$ **7.** $(2x + 1)(x + 2)$

Each pair of factors represents the length and width of a rectangle. Draw an area model and find the area of the rectangle.

8. $(x + 3)(x + 3)$ **9.** $(x + 5)(x + 2)$ **10.** $(3x + 4)(2x + 3)$

Find each product using the FOIL method.

11. $(3x + 1)(x + 3)$ **12.** $(2x + 3)(4x + 1)$ **13.** $(5x + 1)(2x - 3)$

14. $(3x - 5)(2x + 1)$ **15.** $(5x - 2)(3x - 1)$ **16.** $(2x - 1)(3x - 1)$

Mixed Review

Solve each equation for x.

17. $5x + 9 = -18 + 2x$

18. $\dfrac{15}{x} = \dfrac{9}{20}$

19. $x + 30\% x = \$65$

Write and solve an equation for each situation.

20. Taj and Ramen invested $360 in stock. Taj invested $120 more than Ramen. How much did Taj invest?

21. If a rectangle has a length that is 3 cm less than four times its width, and its perimeter is 19 cm, what are its dimensions?

22. John has $15,000 to invest. He invests part of the money in an investment that pays 12% simple interest and the rest in another that pays 8% simple interest. If his annual income from the two investments is $1456, how much did he invest at 12%?

LESSON 10.7 FACTORING TRINOMIALS GEOMETRICALLY

When multiplying two binomials, you are given two factors, and you must find the product.

$$(x + 2)(x + 1) = x^2 + 3x + 2$$

Binomial factors   Trinomial product

When factoring, you do the *opposite*. For example, if given the trinomial $x^2 + 3x + 2$, you can find its factors.

$$x^2 + 3x + 2 = (x + 2)(x + 1)$$

Given trinomial   Binomial factors

Factoring $x^2 + bx + c$

Algeblocks or area diagrams are useful models for factoring trinomials that are written in the form $x^2 + bx + c$. In an area diagram, $x^2 + bx + c$ represents the area of a rectangle. The factors represent the length and width of the rectangle. For example, to find the factors of $x^2 + 5x + 6$, arrange the blocks that represent $x^2 + 5x + 6$ into a rectangle.

The sides of the rectangle, $x + 2$ and $x + 3$, are the factors of $x^2 + 5x + 6$.

ACTIVITY 1   Using Algeblocks to Factor a Trinomial

1. Select the blocks that model the trinomial $x^2 + 4x + 3$.

2. Build a rectangle from your blocks.

3. What are the dimensions of your rectangle?

4. What are the factors of $x^2 + 4x + 3$?

5. Multiply your factors to check your work.

Factoring $ax^2 + bx + c$

You can represent a trinomial such as $2x^2 + 9x + 4$ with a combination of area diagrams.

Area: $2x^2$ Area: $9x$ Area: 4

$$2x^2 + 9x + 4$$

The 15 individual areas form a rectangle.

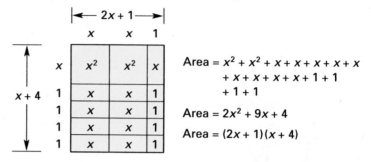

$$\text{Area} = x^2 + x^2 + x + x + x + x + x$$
$$+ x + x + x + x + 1 + 1$$
$$+ 1 + 1$$

$$\text{Area} = 2x^2 + 9x + 4$$

$$\text{Area} = (2x + 1)(x + 4)$$

The length $2x + 1$ and the width $x + 4$ of the rectangle represent the factors of $2x^2 + 9x + 4$.

The model for $2x^2 + 9x + 4$ formed a rectangle. However, some trinomial models such as $x^2 + x + 1$ cannot be arranged in a rectangle. The trinomial $x^2 + x + 1$ is *prime* and *cannot* be factored.

Ongoing Assessment

Use an area diagram to factor $2x^2 + 5x + 3$.

Critical Thinking   Trinomials often contain negative terms. To model the trinomial, some of the areas must be drawn in the second or fourth quadrants. Explain why this is so.

ACTIVITY 2   **Factoring Trinomials with Negative Terms**

Algeblocks Quadrant Mat

1. Use Algeblocks to model the trinomial $x^2 - 5x + 6$.

2. Place the x^2 block in the first quadrant.

3. Place the six unit blocks in the third quadrant.

4. Place the five x-blocks in the second and fourth quadrants.

5. Arrange the blocks to form a rectangle.

6. What are the dimensions of the rectangle?

7. What are the factors of $x^2 - 5x + 6$?

8. Multiply the factors to check your answer.

LESSON ASSESSMENT

1 Can you rearrange your Algeblocks to form different rectangles from the same trinomial? Why or why not?

2 How does using a quadrant map allow you to represent negative values?

Practice and Problem Solving

Use Algeblocks or area diagrams to factor these trinomials.

3. $a^2 + 8a + 7$ **4.** $x^2 - 4x + 3$ **5.** $x^2 - 2x + 1$

6. $x^2 - 8x + 7$ **7.** $a^2 + a - 6$ **8.** $c^2 - 4c + 3$

9. $8 - 6w + w^2$ **10.** $7x^2 + 12x + 5$ **11.** $6x^2 + x - 15$

Mixed Review

As seen from the front, the rafters of the roof of a storage shed form a triangle. The base of the triangle is 12 feet. The perpendicular distance from the peak of the roof to the base is 3 feet.

12. Draw a diagram to model the triangle.

13. If the triangular area formed by the rafters is to be covered with plywood, how many square feet of plywood are needed?

14. If two-thirds of the plywood is painted red, how many square feet of the plywood are painted?

Solve each equation for x.

15. $5x^2 = 125$ **16.** $2x^2 = 3$ **17.** $3x + 6 = x - 18$

From a group of five men and ten women, a committee of four is chosen by drawing names. Find the probability that

18. Four women are chosen.

19. Four men are chosen.

20. Two women are chosen, and then two men are chosen.

LESSON 10.8 FACTORING TRINOMIALS

In Lesson 10.7, you used a geometric model to factor $x^2 + 3x + 2$ into the product of two binomials.

$$x^2 + 3x + 2 = (x + 2)(x + 1)$$

As you know, the factors can be checked by using the Distributive Property or the FOIL method. You can also use the Distributive Property to factor a trinomial.

Factoring $x^2 + bx + c,\ c > 0$

Factoring trinomials involves a step-by-step process and is just like building a model airplane

Factoring $x^2 + 5x + 6$ requires you to find two binomials of the form $(x + m)(x + n)$, whose product is $x^2 + 5x + 6$.

If the Distributive Property is used to multiply $(x + m)(x + n)$, the product is

$$x^2 + mx + nx + mn \quad \text{or} \quad x^2 + (m + n)x + mn$$

Thus, the two numbers m and n must have a sum of 5 and a product of 6. There are four different pairs of integers whose product is 6.

$$1 \cdot 6 \qquad -1 \cdot -6 \qquad 2 \cdot 3 \qquad -2 \cdot -3$$

However, only one pair (2 and 3) has a sum of 5. Consequently, you replace m by 2 and n by 3.

$$x^2 + 5x + 6 = (x + m)(x + n)$$
$$= (x + 2)(x + 3)$$

EXAMPLE 1 Factoring $x^2 - 7x + 12$

Factor $x^2 - 7x + 12$.

SOLUTION

This time, you must find a pair of integers whose product is 12 and whose sum is -7. First, list the integer pairs whose products are 12.

$$\begin{array}{ll} 1 \cdot 12 & -1 \cdot -12 \\ 2 \cdot 6 & -2 \cdot -6 \\ 3 \cdot 4 & -3 \cdot -4 \end{array}$$

The sum of -3 and -4 is -7. Thus,

$$x^2 - 7x + 12 = (x - 3)(x - 4)$$

Always use the FOIL method to check your factors. Do the factors check in this example?

Ongoing Assessment

Factor each trinomial. Check your answer using the FOIL method.

a. $x^2 + 12x + 20$ **b.** $x^2 - 9x + 20$

Factoring $x^2 + bx + c, c < 0$

ACTIVITY   **Factoring $x^2 + bx + c, c < 0$**

1 Explain why you can write the trinomial $x^2 + 2x - 15$ in the form $x^2 + 2x + (-15)$.

2 List the factor pairs for -15.

3 Which pair has a sum of 2?

4 Use the pair you found in Step 3 to make this statement true.

$$x^2 + 2x - 15 = (x + ?)(x + ?)$$

5 Rewrite the statement in the following form:

$$x^2 + 2x - 15 = (x + ?)(x - ?)$$

6 Why must one of the signs in final factored form be negative? How did you choose the correct integer to place after the negative sign?

EXAMPLE 2 Factoring $x^2 - x - 12$

Factor $x^2 - x - 12$.

SOLUTION

When the product is negative, you write one of the factors in the form $(x + ?)$ and the other in the form $(x - ?)$. Since the last term of the trinomial is -12, the factored form of the trinomial is

$$(x + ?)(x - ?)$$

You need to find two integers whose product is -12 and whose sum is -1. First, list the integer pairs whose products are -12.

$$1 \cdot -12 \qquad -1 \cdot 12$$
$$2 \cdot -6 \qquad -2 \cdot 6$$
$$3 \cdot -4 \qquad -3 \cdot 4$$

$3 \times (-4) = -12$
$3 + (-4) = -1$

The sum of 3 and -4 is -1. Thus,

$$x^2 - x - 12 = (x + 3)(x - 4)$$

Ongoing Assessment

Factor each trinomial. Check your answer using the FOIL method.

a. $x^2 + 6x - 16$ **b.** $x^2 - 2x - 3$

Factoring $ax^2 + bx + c$

Factoring a trinomial is a guess and check process. Once you have listed the possible factors, the signs in the trinomial determine which pair to choose. After you write the factors, use the FOIL method to check your answer.

Using guess and check on expressions such as $2x^2 + 13x + 15$ may take considerable time. However, the FOIL method does tell us some things about the possible factors.

If the expression $2x^2 + 13x + 15$ can be factored, the first terms of the binomial factors must be x and $2x$. Since the product of the last two terms is 15 and the middle term is positive, the integer pairs must be

$$1 \cdot 15 \quad \text{or} \quad 3 \cdot 5$$

Here are the four possibilities:

1. $(2x + 1)(x + 15)$ **2.** $(2x + 15)(x + 1)$

3. $(2x + 3)(x + 5)$ **4.** $(2x + 5)(x + 3)$

Since the middle term is +13, the sum of the inner and outer products must be 13. Only possibility **3** works.

$$2x^2 + 13x + 15 = (2x + 3)(x + 5)$$

Factoring expressions such as $2x^2 + 13x + 15$ is relatively easy because 2 is prime. When the numbers in the first and last positions are composite numbers, the task can become very tedious. In Chapter 11, you will learn a formula to help you factor any trinomial expression. Remember, factoring a trinomial is like building an airplane. If you take a step-by-step approach, you will be successful.

LESSON ASSESSMENT

Think and Discuss

In the trinomial $x^2 + bx + c$, what do you know about the factors of c if

1 c is positive and b is negative?

2 c is positive and b is positive?

3 c is negative and b is positive?

4 c is negative and b is negative?

Practice and Problem Solving

Factor each trinomial. Check your answer by using the FOIL method.

5. $x^2 + 5x + 6$ **6.** $x^2 + 7x + 6$ **7.** $x^2 - 5x + 6$

8. $x^2 - x + 6$ **9.** $x^2 + 5x - 6$ **10.** $x^2 - 7x + 6$

11. $5x^2 + 7x + 2$ **12.** $2x^2 + 5x - 12$ **13.** $3x^2 + 14x + 15$

14. $4x^2 + 8x + 3$ **15.** $6x^2 + 7x - 20$ **16.** $6x^2 - 19x + 15$

Mixed Review

Write and solve an equation in one variable for each situation.

17. Reece charges $18 to cut and trim a yard. How much does Reece charge for trimming if he charges $15 to cut the yard?

18. Mary, Pat, and Armond have each earned money for basketball camp. Armond has earned twice as much as Pat, and Mary has earned three times as much as Pat. If the total earned is $210, how much has each person earned?

19. The length of a rectangle is equal to the base of a triangle. The altitude of the triangle is 20 feet, and the height of the rectangle is 25 feet. If the combined areas of the triangle and the rectangle are 280 square feet, what is the length of the rectangle?

Solve each system of equations.

20. $3x + y = 11$ **21.** $2x + 5y = -3$
     $2x - y = 10$ $x - 3y = -7$

Write and solve a system of equations for this problem.
22. A store owner made $2950 during a shirt and pants sale. The shirts sold for $40, and the pants sold for $65. There were 5 more pants sold than shirts. How many shirts were sold?

LESSON 10.9 SPECIAL PRODUCTS AND FACTORS

The Sum and Difference of Two Squares

The binomial $a + b$ is the sum of two terms. The binomial $a - b$ is the difference of the same two terms. The product of the sum and difference is represented as

$$(a + b)(a - b)$$

sum • difference

Simplify $(9 + 1)(9 - 1)$ using the FOIL method.

$$(9 + 1)(9 - 1) = 81 - 9 + 9 - 1$$
$$= 81 + 0 - 1$$
$$= 9^2 - 1^2$$

Notice that:

- the sum of the inner product and the outer product is zero.
- the final product is the difference of two squares.

The multiplication $(9 + 1)(9 - 1)$ results in a **special product**:

$$9^2 - 1^2$$

Special products enable you to use a shortcut to write the answer for certain multiplication and factoring problems.

Examine the procedure for multiplying $(a + b)(a - b)$. Again, you see that the inner product (ab) and outer product ($-ab$) cancel because they add to zero. The product of the sum and difference of two numbers is the difference of their squares.

$$(a + b)(a - b) = a^2 - ab + ab - $$
$$= a^2 + 0 - b^2$$

ACTIVITY 1   Factoring the Difference of Two Squares

1 Factor $x^2 - 4$ using Algeblocks.

2 Check your answer using multiplication.

3 Factor $x^2 - 25$ using Algeblocks.

4 Check your answer using multiplication.

5 The terms x^2 and 25 are perfect squares. How can you use square roots to factor the expression $x^2 - 25$?

6 Explain how to factor the difference of two perfect squares.

7 Factor $a^2 - 36$.

In Activity 1, you found a pattern for factoring the difference of two squares. For example, $x^2 - 100 = x^2 - 10^2 = (x + 10)(x - 10)$.

> ### Factoring the Difference of Two Squares
> Since $(a + b)(a - b) = a^2 - b^2$,
> $$a^2 - b^2 = (a + b)(a - b)$$

EXAMPLE 1 The Product of a Sum and Difference

a. Multiply $(2x + 3y)(2x - 3y)$.

b. Factor $16s^2 - 49t^2$.

SOLUTION

a. Use the shortcut.

Square the first term $2x$, $(2x)^2 = 4x^2$

Square the second term $3y$, $(3y)^2 = 9y^2$

Write the difference of the squares, $4x^2 - 9y^2$

Thus,

$$(2x + 3y)(2x - 3y) = 4x^2 - 9y^2$$

b. Use the shortcut. The first term $16s^2$ is a perfect square, $[(4s)^2]$. The second term $49t^2$ is also a perfect square $[(7t)^2]$. Thus,

$$16s^2 - 49t^2 = (4s + 7t)(4s - 7t)$$

Ongoing Assessment

a. Multiply $(m - 3n)(m + 3n)$.   **b.** Factor $16 - 81w^2$.

Critical Thinking   Why is $x^2 + 9$ a prime expression?

Perfect Square Trinomials

When you square a number, you use the number as a factor two times. Thus, $3^2 = 3 \cdot 3$, or 9. In general, the square of x is x^2. What does it mean to square a binomial?

ACTIVITY 2   Squaring a Binomial

1 Find each product using Algeblocks or an area model.

    a. $(x + 3)(x + 3)$ **b.** $(x - 3)(x - 3)$

    c. $(x - 4)(x - 4)$ **d.** $(x + 4)(x + 4)$

2 Find each product using the FOIL method.

    a. $(2 + m)(2 + m)$ **b.** $(2 - m)(2 - m)$

    c. $(m - n)(m - n)$ **d.** $(m + n)(m + n)$

3 Complete the equation.

$$(a + b)^2 = (\underline{\ ?\ } + \underline{\ ?\ })(\underline{\ ?\ } + \underline{\ ?\ }) = \underline{\ ?\ } + \underline{\ ?\ } + \underline{\ ?\ }$$

4 What pattern do you observe when you square a binomial?

A **perfect square trinomial** is a special trinomial that can be factored into two identical binomials called a **binomial square**.

$$\underbrace{x^2 + 6x + 9}_{\substack{\text{perfect square}\\\text{trinomial}}} = \underbrace{(x + 3)(x + 3)}_{\substack{\text{two identical}\\\text{binomials}}} = \underbrace{(x + 3)^2}_{\substack{\text{binomial}\\\text{square}}}$$

EXAMPLE 2   Factoring $4x^2 + 4x + 1$

Is the polynomial $4x^2 + 4x + 1$ a perfect square trinomial?

SOLUTION

If $4x^2 + 4x + 1$ is a perfect square trinomial, then it factors into a binomial square.

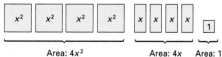

When you arrange the areas into the shape of a rectangle, the rectangle is a square.

In Activity 2, you observed a pattern that is used to square a binomial or factor a perfect square trinomial such as $4x^2 + 4x + 1$. From this pattern, note these three questions, which will help determine if a trinomial is a perfect square trinomial.

- Are the first and last terms positive? In the trinomial $4x^2 + 4x + 1$, both $4x^2$ and 1 are positive.

- Are the first and last terms perfect squares? $4x^2 = (2x)^2$ and $1 = 1^2$.

- Find the square roots of the first and last terms. Is twice the product of the square roots equal to the middle term?

$$2(2x)(1) = 4x$$

If a trinomial satisfies these three conditions, it is a perfect square.

When a binomial is squared, the result is a perfect square trinomial. When a perfect square trinomial is factored, the result is a binomial square.

Perfect Square Trinomials
$$(a + b)^2 = a^2 + 2ab + b^2$$
$$(a - b)^2 = a^2 - 2ab + b^2$$

Ongoing Assessment

Which of the following expressions are perfect square trinomials? Explain your answer.

a. $x^2 - 14x + 49$ **b.** $x^2 + 5x + 25$ **c.** $x^2 + 12x - 36$

EXAMPLE 3 Factoring a Perfect Square Trinomial

Factor $x^2 - 10x + 25$.

SOLUTION

The first term x^2 is a perfect square.

The last term is a perfect square $(-5)^2 = 25$.

The middle term is $2(x)(-5)$.

Thus, $x^2 - 10x + 25 = (x - 5)(x - 5) = (x - 5)^2$.

a. Multiply $(2t + r)^2$.   **b.** Factor $4x^2 + 12xy + 9y^2$.

A word of caution about squaring a binomial: *Do not forget the middle term*. A common error is to think that $(a + b)^2$ equals $a^2 + b^2$.

Substitute 1 for a and 2 for b, and you will see that this is not the case.

WORKPLACE COMMUNICATION

Show how to obtain the approximation: (a) expand the square of the binomial in the equation for the exact volume, and then (b) multiply by π using the Distributive Property. What assumption is required? Compare the exact and approximate volumes for $r = 10$ mm, $a = 0.1$ mm, and $L = 1$m.

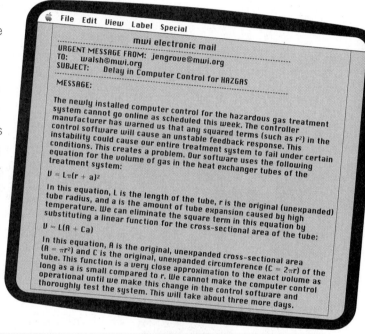

File Edit View Label Special

mwi electronic mail

URGENT MESSAGE FROM: jengrove@mwi.org
TO: walsh@mwi.org
SUBJECT: Delay in Computer Control for HAZGAS

MESSAGE:

The newly installed computer control for the hazardous gas treatment system cannot go online as scheduled this week. The controller manufacturer has warned us that any squared terms (such as r^2) in the control software will cause an unstable feedback response. This instability could cause our entire treatment system to fail under certain conditions. This creates a problem. Our software uses the following equation for the volume of gas in the heat exchanger tubes of the treatment system:

$$V = L\pi(r + a)^2$$

In this equation, L is the length of the tube, r is the original (unexpanded) tube radius, and a is the amount of tube expansion caused by high temperature. We can eliminate the square term in this equation by substituting a linear function for the cross-sectional area of the tube:

$$V \approx L(A + Ca)$$

In this equation, A is the original, unexpanded cross-sectional area ($A = \pi r^2$) and C is the original, unexpanded circumference ($C = 2\pi r$) of the tube. This function is a very close approximation to the exact volume as long as a is small compared to r. We cannot make the computer control operational until we make this change in the control software and thoroughly test the system. This will take about three more days.

LESSON ASSESSMENT

Think and Discuss

1 What happens to the middle term when you multiply the sum and difference of the same two numbers or terms?

2 How do you determine the middle term when you square a binomial?

3 What is the general rule or shortcut for finding the product of the sum and difference of the same terms?

4 What is the general rule or shortcut for squaring a binomial?

5 How can you use the general rules for multiplying in these special cases to find the factors of special trinomials?

Multiply.

6. $(x + 1)(x - 1)$ **7.** $(x - 3)(x + 3)$ **8.** $(2x - 5)(2x + 5)$

9. $(x + y)(x + y)$ **10.** $(2x - 1)(2x - 1)$ **11.** $(x - 3)(x + 3)$

12. $(a - 4)^2$ **13.** $(2y + 3)^2$ **14.** $(4x - 3y)(4x + 3y)$

Factor.

15. $x^2 - 25$ **16.** $4y^2 - 9$ **17.** $9 - 16y^2$

18. $x^2 - 6x + 9$ **19.** $y^2 - 18y + 81$ **20.** $9x^2 + 24x + 16$

21. $y^2 - 2y + 1$ **22.** $y^2 - 1$ **23.** $x^2 - 2ax + a^2$

Mixed Review

24. Solve $F = \frac{9}{5}C + 32$ for C. **25.** Solve $P = 2a + 2b$ for b.

A cartographer has located three points on a map. The points are designated as $A(-3,0)$, $B(2,1)$, and $C(4,-1)$.

26. Connect the points in order from A to B to C. What geometric figure is formed?

27. Find the equation of the line that passes through points A and B.

28. Find the equation of the line that passes through Point C which is parallel to the x-axis.

29. Find the equation of the line that passes through point A which is perpendicular to the x-axis.

Solve each system of equations.

30. $x - y = 1$ **31.** $2x - 5y = 9$ **32.** $\frac{1}{2}x - \frac{1}{3}y = 1$

$2x + y = 0$ $2x - 2y = 3$ $\frac{1}{3}x + \frac{1}{2}y = 5$

MATH LAB

Activity 1: Factoring Trinomials with Math Tiles

Equipment Cardboard, 7.5 inches by 4 inches
Scissors

Problem Statement

Use the template to create a set of math tiles.

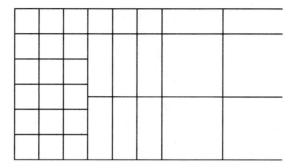

Mark the large squares with an x^2. Mark the long rectangles with an x. Mark the small squares with a 1. After you cut the tiles apart, you will have four x^2-tiles, eight x-tiles, and twenty-one 1-tiles.

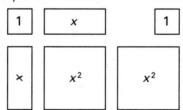

You will use the math tiles to model trinomial factorization. You will start with the tiles that represent each term of the trinomial and use those tiles to try and form a rectangle.

Procedure

a The polynomial $x^2 + 5x + 6$ is modeled by one x^2-tile, five x-tiles, and six 1-tiles.

Start with the x^2-tile and add on the areas until you can make a rectangle. Always line up the long sides of the x-tiles against each other.

Like This   Not Like This

b Now complete the rectangle.

c When you have made a rectangular area with the tiles, draw a picture of your rectangle and record the dimensions along its base and side. These dimensions represent the factors of the polynomial. What are the dimensions of the rectangle you have just made? What are the factors of $x^2 + 5x + 6$?

d As you work with more examples, it will become easier to form the rectangles. If it is not possible to form a rectangle with the tiles, the trinomial is prime and cannot be factored. Explain why $x^2 + x + 2$ is prime.

e Use tiles to find the factors for each trinomial.

1. $x^2 + 5x + 6$

2. $x^2 + 7x + 6$

3. $x^2 + 7x + 12$

4. $x^2 + 8x + 12$

5. $x^2 + 6x + 9$

6. $x^2 + 4$

f Based on your observations in Step **e**, explain how you would factor $x^2 + bx + 8$.

g Use tiles to find the factors of each trinomial.

7. $2x^2 + 7x + 3$ **8.** $2x^2 + 5x + 3$

h How can you make a set of tiles to factor $x^2 + 2xy + y^2$? Explain how to use your tiles to show that $x^2 + 2xy + y^2 = (x + y)(x + y)$.

Activity 2: Listing Prime Numbers

Equipment   Set of 30 small objects, such as pennies, beans, or math tiles

Problem Statement

You will discover if a number is *prime* by arranging objects into rectangular arrays.

Procedure

a   For each of the numbers *n* greater than 1 and less than 30, count out *n* objects.

b   Find all the possible rectangular arrangements, called **arrays**, of *n* objects. Here is an example for the number 6:

1 by 6	6 by 1	2 by 3	3 by 2
# # # # # #	#	# # #	# #
	#	# # #	# #
	#		# #
	#		
	#		
	#		

c   For each array of *n* objects, record the array with the number that represents the length and the width next to the array.

d   List the numbers from 2 through 29.

e   Circle any number *n* on your list that can only be represented by an array of 1 by *n* and *n* by 1. The numbers that you have circled are the prime numbers less than 30.

f   Compare your list of prime numbers with others in your class. Did all of you get the same results?

g   Write a definition of a prime number.

MATH APPLICATIONS

The applications that follow are like the ones you will encounter in many workplaces. Use the mathematics you have learned in this chapter to solve the problems.

Wherever possible, use your calculator to solve the problems that require numerical answers.

A formula to find the sum of the first n whole numbers $1 + 2 + 3 + ... + n$ is

$$\text{Sum} = \frac{n^2 + n}{2}$$

Thus, for $n = 6$, the formula should give the sum of the first six whole numbers, or $1 + 2 + 3 + 4 + 5 + 6 = 21$.

1 With a calculator, it might be easier to evaluate this formula without having to compute the n^2 term. Identify the common monomial term in the numerator of the formula and factor out this term to obtain a different version that does not have the squared term.

2 Test the formula, using a relatively small value for n, to show that it works.

3 What is the sum of the first 200 whole numbers?

4 What is the sum of the first 1000 whole numbers?

Multiply each pair of binomials below using the FOIL method.

5 $(x + 5)(x + 2)$

6 $(x - 3)(3x + 8)$

7 $(4x + 1)(2x - 5)$

8 $(7x - 2)(x - 4)$

Match each binomial product with the correct "area" on the right that geometrically represents it. Multiply the products for each and verify that the areas match the polynomial.

9 $(2x + 1)(x + 2)$

a.

10 $(x + 3)(x + 1)$

b.

11 $(x + 1)(2x + 1)$

c.

Factor each of the monomials below into its prime factors.

12 12   **13** 94   **14** $243x^2$ **15** $180x^5$

Factor the GCF from each polynomial.

16 $3x^2 + 18x$ **17** $12x^3 - 30x^2$ **18** $105x^4 + 21x^3 - 63x^2$

Use the Distributive Property to find each product.

19 $(x + 7)(x^2 + 6x - 9)$ **20** $(2x - 3)(3x^2 - 1)$

21 $(2x + 5)(9x^2)$ **22** $(x - 5)(x^2 - x)$

To factor a trinomial, one of the skills needed is the ability to identify pairs of numbers that add to give a certain sum and multiply to yield a given product. For example, 3 and 2 add to 5, and when multiplied give 6. Copy the table and complete it by providing a pair of numbers whose sum is the number in the first column and whose product is the number in the second column. (The first row is completed for you.)

Sum ($a + b$)	Product (ab)	a	b
5	6	3	2
23 2	1	?	?
24 8	15	?	?
25 11	10	?	?
26 14	48	?	?
27 −4	4	?	?
28 −6	−7	?	?
29 2	−8	?	?
30 9	−52	?	?
31 0	−9	?	?

Factor each trinomial. If the trinomial cannot be factored, then indicate that it is *prime*.

32 $x^2 + 6x + 5$ **33** $x^2 + 14x + 45$

34 $x^2 - x - 20$ **35** $x^2 + x + 16$

36 $x^2 + x - 2$ **37** $x^2 - 5x - 6$

Factor each trinomial below or state that it is *prime*.

38 $2x^2 + 11x + 5$

39 $3x^2 - 11x - 4$

40 $4x^2 - 15x - 4$

41 $6x^2 - 5x - 4$

Identify which of the following trinomials are considered "special products" and factor them. (You need not factor those that are not "special products.")

42 $x^2 - 4$

43 $x^2 + 6x + 9$

44 $4x^2 - 2x + 1$

45 $4x^2 + 1$

46 $x^2 - 49$

47 $x^2 - 8x + 16$

AGRICULTURE & AGRIBUSINESS

You are constructing a large rack to hold small pots while the seeds in them germinate. You plan to have approximately 500 pots in the rack. You would also like the rack to be as close to "square" as possible. The rack should have an equal number of pots in each row and an equal number of pots in each "column" (although possibly different from the number in each row). What are some possible numbers of pots per row and pots per column that would give you a "nearly square" layout for the following total number of pots?

48 480

49 500

50 520

51 Is there some total number of pots near 500 that you can use to create a perfectly square layout? If so, what is it?

BUSINESS & MARKETING

You are considering packaging small objects into large rectangular shaped cartons for shipping. The cartons should hold no more than 6 objects per row (length), be no more than 6 objects high (height), and be no more than six objects deep (depth). You would like to package no less than 100 objects per carton.

52 What are all your choices of total number of objects per carton? (Empty spaces in the carton are not allowed.)

A quilt is to be made of patchwork squares. The quilt has a length of *L* squares and a width of *W* squares. Although many different sizes are possible, you would like the quilt's length-to-width ratio to be 5 to 3. Thus, for a quilt of "size" *S*, if the quilt has a length of *L* squares equal to $5S$, the acceptable width would be *W* squares equal to $3S$.

53 Write an equation that gives the total number of squares *T* in the quilt as a function of the size *S*.

54 The quilt can be made slightly larger by adding a row of squares along the length and a row along the width. Modify your equation to show the total number of squares if the quilt is $L + 1$ squares long and $W + 1$ squares wide. Write your expression as two factors.

55 Multiply the two factors using the FOIL technique to obtain a polynomial expression.

The cost to produce round metal washers is directly related to the amount of metal used. And the amount of metal used is directly related to the area of the washer, as shown in the shaded region.

56 What is the area of the large disc of radius *R* before the middle hole is stamped out?

57 What is the area of the middle hole of radius *r*?

58 Write a formula for the area of the washer that has an outer radius *R* and a hole of radius *r*, using the formulas from Exercises 56 and 57. Simplify the formula by factoring out all common monomial terms.

59 You should recognize one of the factors in your formula as a "special product." Show your formula for the area of a washer in yet another form by rewriting the "special product" as two factors.

In the previous exercises, you found a formula that gives the area of metal used in a washer. When making this washer, a certain manufacturer will start with a square piece of metal and make the two circular stamps.

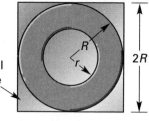

The length of the square's side will be equal to the diameter of the washer, or $2R$. Thus, the material outside the diameter of the washer as well as the hole from the middle becomes "waste."

Metal Plate

60 Write a formula for the area of the square piece of metal that will be used to stamp out a washer with a diameter of $2R$.

61 Create a formula for the area of metal waste from the production of a single washer of outer radius R and inner radius r. You can do this by subtracting the area of the washer (see Exercise 60) from the area of the square piece of metal. Simplify the formula by factoring out the GCF.

62 Use your formula from Exercise 58 to determine how many square inches of metal are used in creating a washer that has an outer radius of $\frac{3}{4}$ inch and an inner radius of $\frac{1}{2}$ inch.

63 How many square inches of metal would be wasted in creating such a washer (using your formula from Exercise 61)?

64 Comparing the metal wasted to the area of the initial metal square, what percent of metal is "wasted" in creating one of these metal washers?

A rectangular piece of metal has a length that is 5 cm greater than its width W in centimeters. As part of an assembly, the plate must have four square sections removed from it, each having a width of $\frac{1}{4}W$.

65 Write a formula for the area of the metal plate that remains after the corner pieces are removed.

66 Simplify the formula from Exercise 65 by factoring out the GCF.

67 Use your formula to find the area of the metal plate that remains when the width is chosen as $W = 24$ cm.

CHAPTER 10 ASSESSMENT

Skills

1. Find the sum. $(5x^2 + 3x - 9) + (2x^2 - 2x + 3)$

2. Find the difference. $(3x^2 - 5x + 2) - (4x^2 + 3x - 4)$

3. What is the greatest common factor of 36 and 90?

4. What expression represents the shaded area between the circle and the square? Write the expression in factored form.

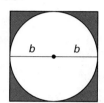

Find each product.

5. $3x^2(x^2 - 3x + 8)$ 6. $(x + 4)(x - 1)$ 7. $(x - 3)^2$

Simplify each expression.

8. $(5x^2)^3$ 9. $(3x^2)(2x^5)$ 10. $(4.1 \times 10^2)^3$

Factor each expression.

11. $x^2 + 2x - 15$ 12. $4x^2 - 9$ 13. $4x^2 - 4x + 1$

Applications

14. The length of a rectangular floor is 3 meters longer than its width. Write an expression that represents the area of the rectangle.

15. A square playground's area is represented by the trinomial $x^2 - 10x + 25$. Write an expression using the variable x that represents the perimeter of the playground.

CHAPTER 11

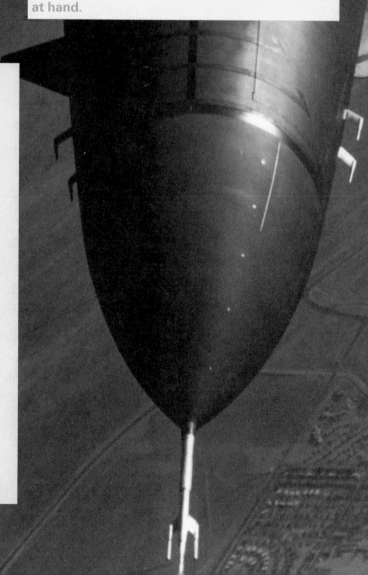

WHY SHOULD I LEARN THIS?

There are many ways to solve a problem. Some are more efficient than others. To design a supersonic aircraft, an engineer must select a design method that completes the job quickly and efficiently. In this chapter, you will learn several methods to solve quadratic equations. Master these methods and learn to select the one that most efficiently solves the problem at hand.

QUADRATIC FUNCTIONS

OBJECTIVES

1. Solve quadratic equations by graphing.
2. Solve quadratic equations by factoring.
3. Solve quadratic equations with the quadratic formula.
4. Use graphing, factoring, and the quadratic formula to solve practical problems involving quadratic equations.

In Chapter 6 you graphed several types of nonlinear equations. Now you will revisit one particular type of nonlinear equation. In this chapter, you will examine how the quadratic equation and its graph, the parabola, are used in science and industry.

The parabola has many applications in the modern world of technology. Supersonic airplanes are parabolic in shape to reduce air resistance. Flashlights and satellite dish antennas have reflectors shaped like a parabola. The parabolic reflector in a flashlight reflects the light from a tiny bulb, sending out a beam of nearly parallel rays of light. The parabolic reflector of a satellite dish reflects incoming radio signals and focuses them on a receiver unit. Compare the shape of the parabola drawn on the x- and y-axes with the drawings of the flashlight and the satellite dish below. Can you find the parabolic reflectors?

As you watch the video for this chapter, look for graphs that represent parabolas.

LESSON 11.1 GRAPHING QUADRATIC EQUATIONS

Suppose an arrow is fired vertically into the air with an initial speed of 112 feet per second. The height (h) of the arrow in feet is a function of the time (t) in seconds. You can model the relationship between the height and the time of the arrow's flight with the following **quadratic equation**.

$$h = -16t^2 + 112t$$

To graph $h = -16t^2 + 112t$, make a table of ordered pairs (t, h). Plot the points and connect them with a smooth curve. The parabola shows how the arrow rises until it reaches its maximum height. Then it falls back to Earth.

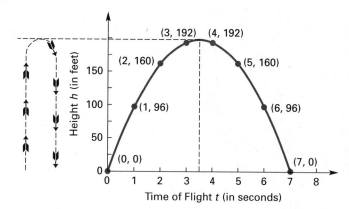

ACTIVITY 1   Maximum Height

The vertical dashed line is drawn at the maximum height of the arrow, which is the vertex of the parabola. The maximum height occurs when t is 3.5 seconds.

1 What is the total time of flight of the arrow?

2 Use the graph to estimate the maximum height of the arrow.

3 Use the equation $h = -16t^2 + 112t$ to calculate the actual value for the maximum height of the arrow.

Juanita is a pitcher for a softball team. Her coach measured the speed of her fastball at 64 feet per second. If Juanita throws the ball upward with an initial speed of 64 feet per second, the following quadratic equation models the relationship between the height (h) of the ball and the time (t) of travel.

$$h = -16t^2 + 64t$$

Critical Thinking Which coefficients in the last two equations represent the initial speeds of the arrow and the softball?

The height of the softball, like the height of the arrow, is a function of time. Thus, you can graph the softball's height on the same set of axes as the arrow's height.

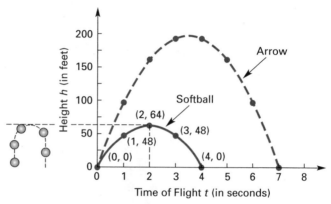

ACTIVITY 2   Time in the Air

Examine the graph of $h = -16t^2 + 64t$.

1 What is the maximum height the softball reaches?

2 How many seconds is the ball in the air?

In Chapter 6, you saw how the coefficients a and c in $y = ax^2 + c$ altered the graph of the parent quadratic function $y = x^2$.

Changing the sign of a reflects the graph across the x-axis.

$$y = x^2$$

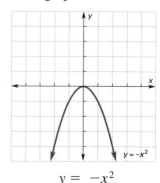

$$y = -x^2$$

Increasing *a* shrinks the graph. Decreasing *a* stretches the graph.

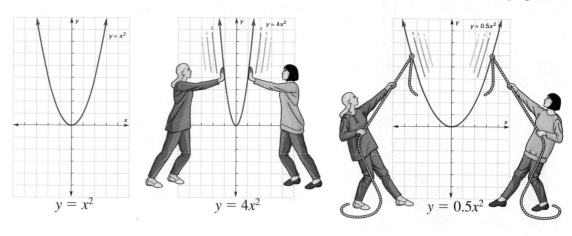

$$y = x^2$$

$$y = 4x^2$$

$$y = 0.5x^2$$

Adding *c* to the parent function translates the graph vertically.

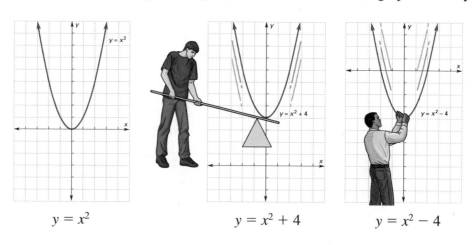

$$y = x^2$$

$$y = x^2 + 4$$

$$y = x^2 - 4$$

If a constant is added or subtracted from the variable before it is squared, the graph is translated horizontally.

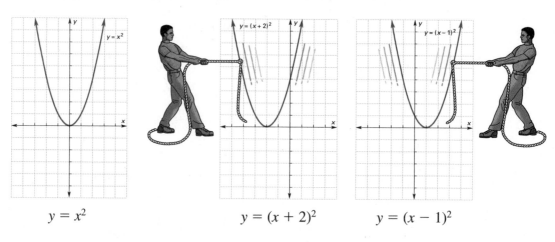

$$y = x^2$$

$$y = (x + 2)^2$$

$$y = (x - 1)^2$$

Reflections, stretching and shrink
vertical and horizontal translations a

Transformation

A *transformation* is a rule that tells ho
related to its parent function.

EXAMPLE Transforming a Parabola

Explain how transformations are used to graph $y = (x - 2)^2 + 1$.

SOLUTION

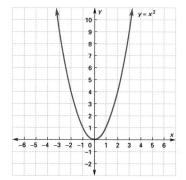

Start with the
parent function.

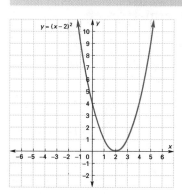

Move the vertex
two units to the right.

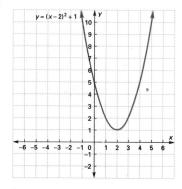

Move the vertex
one unit up.

Critical Thinking Explain how to find the coordinates of the
new vertex when $y = x^2$ is moved to $y = (x - m)^2 + n$?

You know that changing the coefficients of the terms in a quadratic
equation has an effect on the graph of the equation. In the following
lessons, these coefficients will play a key role in defining the
characteristics of a quadratic function.

...equation $h = -16t^2 + 250t$ represent the flight of a ...ocket.

...how to use the graph of the equation to find the ...m height of the rocket.

...how to find the total time the rocket is in the air.

...ith the parent function $y = x^2$. How would you write ...v equation for each of the following transformations? ...e coordinates of the new vertex.

3 The parent graph is translated three units to the left.

4 The parent graph is translated two units up.

5 The parent graph is translated one unit to the right and 1 unit down.

Practice and Problem Solving

Graph each function. Describe the transformation of the parent graph. Give the coordinates of the vertex of the transformed graph.

6. $y = -2x^2$ **7.** $y = 3x^2 + 5$ **8.** $y = -(x + 3)^2$

9. $y = (x - 2)^2 + 1$ **10.** $y = \frac{1}{2}x^2 - 3$ **11.** $y = -\frac{1}{2}(x + 1)^2 + 2$

Mixed Review

The formula for finding compound interest is $A = p(1 + r)^n$.

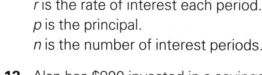

A is the amount of the new balance.
r is the rate of interest each period.
p is the principal.
n is the number of interest periods.

12. Alan has $800 invested in a savings account. The account pays 6% interest compounded yearly. How much will the balance be after four years?

13. Shondra has $800 invested in a savings account. The account pays 8% interest compounded yearly. How much more interest will Shondra earn in four years than Alan?

Shondra
Alan
8%
6%

ing (also called dilations), and
e called **transformations.**

v a function is

14. State Bank compounds interest monthly. City Bank compounds interest semiannually. Each bank pays an annual rate of 6%. How much more interest will you earn if you invest $1000 in State Bank for one year? (*Hint:* The interest rate for each period is the annual rate divided by the number of interest periods.)

Write and solve an equation for each situation.

15. The side of a large square is 2 inches longer than the side of a small square. The sum of the perimeters of the two squares is 40 inches. What is the length of each side of the small square?

16. Raul receives a bonus of 5% of his salary when he meets his sales goal. One week Raul made his goal and earned a total of $256. How much is his bonus?

17. You have three test scores of 85, 90, and 93. What is the lowest score you can make on your fourth test and still have a test average of 90?

Two cars are located on a grid at the coordinates $M(-3, -2)$ and $N(1, 4)$.

18. What is the slope of line *MN*?

19. What is the equation of line *MN*?

20. Another car is located at point $P(2, 2)$. What is the equation of the line passing through point *P* and parallel to line *MN*?

LESSON 11.2 THE QUADRATIC FUNCTION

In Lesson 11.1, you graphed the flight of a softball thrown upward by Juanita. The softball had an initial speed of 64 feet per second, and you used the quadratic equation $h = -16t^2 + 64t$ to model the flight. Since at time $t = 0$, the height is $h = 0$, this equation models a softball starting its flight at ground level.

Suppose a person stands on a 75-foot cliff and throws a rock with the same upward speed as Juanita's fastball. You need a new equation to show the effect of moving the initial starting height from the ground to the 75-foot cliff. This new quadratic equation is

$$h = -16t^2 + 64t + 75$$

You can use the same coordinate axes to graph the equations for the height of the arrow, the softball, and the rock.

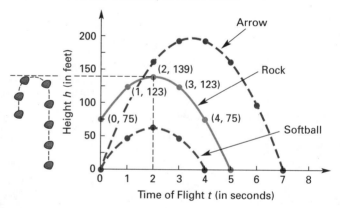

Critical Thinking   What is the maximum height reached by the rock? What is the approximate flight time of the rock from release to impact with the ground?

The graphs representing the softball and the rock are different because at any given moment, the rock parabola is "higher" than the softball parabola. If you slide the rock parabola vertically downward 75 units on the y-axis, it fits right over the softball parabola, point for point. Two figures that match point for point are **congruent**.

The softball and the rock equations are represented by congruent parabolas. They are congruent because the coefficients of t^2 and t in each equation are identical.

Critical Thinking Use the language of transformations to describe the graphs for both the rock and the softball in terms of the parent function $y = t^2$.

An equation written in the form

$$y = ax^2 + bx + c$$

is a *quadratic equation written in standard form*. The heights of the arrow, the softball, and the rock are described by quadratic equations.

$$
\begin{aligned}
y &= ax^2 + bx + c \\
\text{arrow} \longrightarrow \quad h &= -16t^2 + 112t + 0 \\
\text{softball} \longrightarrow \quad h &= -16t^2 + 64t + 0 \\
\text{rock} \longrightarrow \quad h &= \underbrace{-16t^2}_{} + \underbrace{64t}_{} + \underbrace{75}_{}
\end{aligned}
$$

constant term

linear term

quadratic term

A quadratic equation contains three terms. The quadratic term is degree two, the linear term is degree one, and the constant term is a number with no variable.

EXAMPLE 1 Graphing a Quadratic Function

Graph the quadratic equation $y = x^2 + x - 12$.

SOLUTION

The equation $y = x^2 + x - 12$ defines a relationship between x and y. Thus, y is a function of x. The equation written in *functional notation* is

$$f(x) = x^2 + x - 12$$

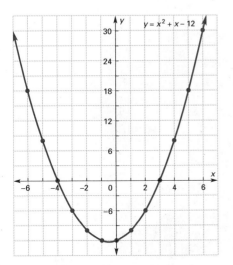

$y = x^2 + x - 12$

x	y
-6	18
-5	8
-4	0
-3	-6
-2	-10
-1	-12
0	-12
1	-10
2	-6
3	0
4	8
5	18
6	30

Substitute integer values for x from -6 to 6 ($-6 \le x \le 6$) to determine the corresponding values of y. The values of x from -6 to 6 are real numbers in the domain of f. The corresponding values of y are in the range of f. Complete a table of ordered pairs, graph the points, and draw the graph. The graph of a quadratic function is *always* a parabola.

Graph the quadratic function $h(m) = m^2 - m - 2$.

Look at the table and graph in Example 1. Where does the graph cross the x-axis? What are the approximate coordinates of the minimum point on the graph? When using a graphics calculator, use the ZOOM key to find a more accurate answer.

The graph crosses the x-axis at $(-4, 0)$ and $(3, 0)$. Since the x-intercepts result when $y = 0$, you have solved the equation

$$x^2 + x - 12 = 0$$

The solutions, also called the **roots,** of the equation $x^2 + x - 12$ are -4 and 3. One way to check your graphing results is to substitute -4 and 3 for x in the function

$$f(x) = x^2 + x - 12$$

$$f(-4) = (-4)^2 + (-4) - 12$$
$$= 0$$

$$f(3) = (3)^2 + (3) - 12$$
$$= 0$$

The Zeros of a Quadratic Function

Since $f(-4) = 0$ and $f(3) = 0$, -4 and 3 are also called the **zeros** or **roots of the function** $f(x) = x^2 + x - 12$.

> ### Zeros or Roots of a Function
> If f is a function and $f(a) = 0$,
>
> then a is a zero or root of f.

Look again at the graph of $f(x) = x^2 + x - 12$. The graph indicates that the vertex, in this case the minimum point, is at $(-0.5, -12.25)$. The vertex is on a vertical line midway between the x-intercepts of the parabola. Since this is true for all parabolas, you can use an algebraic technique to find the minimum or maximum value.

EXAMPLE 2 Finding a Minimum Value

Find the minimum value of $f(x) = x^2 + x - 12$ and the coordinates of the vertex.

SOLUTION

The minimum value of f is the y-value of the vertex. Since the vertex is midway between $(-4,0)$ and $(3,0)$, it must have a first coordinate of $\frac{-4 + 3}{2}$, or -0.5. Its y-coordinate, $f(-0.5)$, is the minimum value.

$$f(-0.5) = (-0.5)^2 + (-0.5) - 12 = 0.25 - (0.5) - 12 = -12.25$$

The vertex of the parabola for f is at $(-0.5, -12.25)$. The minimum value is -12.25.

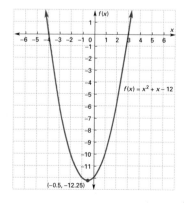

The line that passes through the vertex and is parallel to the y-axis is called the **axis of symmetry.** The axis of symmetry for the parabola in Example 2 is given by the equation $x = -0.5$.

Find the axis of symmetry, the vertex, and the minimum value for the function $f(x) = x^2 - 5x + 6$.

The graphs of some quadratic functions touch the x-axis at only one point. In this case, there is only one root. Consider the equation

$f(x) = x^2 - 6x + 9$

$f(3) = 3^2 - 6(3) + 9$

$f(3) = 0$

The root of the equation is 3.

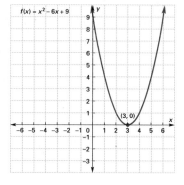

In some instances, the graph does not cross the x-axis at all.

Consider the function $y = f(x) = x^2 + 2x + 6$

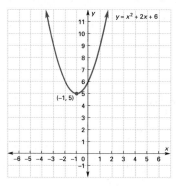

There is no real number that is a solution of this equation when $y = 0$. Thus, there are no real roots.

Approximating Zeros

You can use a graphics calculator to find the zeros of a quadratic function.

EXAMPLE 3 Using a Graphics Calculator

Use a graphics calculator to find the zeros of $f(x) = x^2 + 3x + 1$.

SOLUTION

Graph $y = x^2 + 3x + 1$ on your calculator. Use the ⟨ZOOM⟩ and ⟨TRACE⟩ keys to find the coordinates where the graph intersects the x-axis. These x-coordinates are the zeros. Since the zeros are not integers, you can approximate the zeros by rounding to the nearest thousandth.

$$x \approx -2.618 \text{ and } x \approx -0.382$$

The zeros of the function are about -2.618 and -0.382.

LESSON ASSESSMENT

Think and Discuss

Use $f(x) = x^2 - x - 6$ for 1–5.

1 Explain how to graph f.

2 Explain how to find the zeros of f.

3 Explain how to find $f(3)$.

4 How do you find the maximum or minimum of f?

5 How do you find the vertex and the axis of symmetry for the graph of f?

Practice and Problem Solving

Use a graph to find the x-intercepts and the vertex.

6. $y = x^2 - 8x + 15$ **7.** $y = x^2 + 6x + 8$

8. $y = x^2 + x - 2$ **9.** $y = x^2 + 2x + 1$

10. $y = x^2 + 5x + 6$ **11.** $y = x^2 + 3x - 10$

Use a graphics calculator to find the roots of each function.

12. $y = x^2 - 2x + 1$ **13.** $y = x^2 - 4x - 5$

14. $y = x^2 - 1$ **15.** $y = x^2 + 8x + 3$

16. $y = x^2 + 4x - 6$ **17.** $y = x^2 - 3.8x - 0.8$

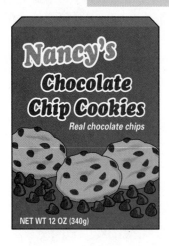

The profit *P* for Nancy's Cookie Company is determined by subtracting overhead *O* from income *I*.

18. Write a formula to find the profit for Nancy's company.

19. Find the profit when the income is $2500 and the overhead is $1600.

20. Find the income when the profit is $500 and the overhead is $1200.

21. A rectangular box has a length of 20 centimeters, a width of 15 centimeters, and a depth of 8 centimeters. Suppose the length is increased by 2 centimeters and the depth is decreased by 2 centimeters. By how much is the volume of the new box increased or decreased?

Solve each equation.

22. $7a - 9 = 2a + 11$ **23.** $\dfrac{2x}{8} = \dfrac{12}{24}$ **24.** $80\% \ m + m = 126$

25. Find the slope and *y*-intercept of $x + y = 6$.

26. Find the slope and *y*-intercept of $y - 3x = 6$.

27. Solve this system of equations: $x + y = 6$
$y - 3x = 6$

Solve using a system of equations.

28. Roberto sells caps for $20 and T-shirts for $40. On Saturday Roberto sold a total of 12 caps and T-shirts. He received $320. How many caps did Roberto sell?

LESSON 11.3 SOLVING A QUADRATIC EQUATION BY FACTORING

Examine the graph of $y = x^2 - 3x - 10$.

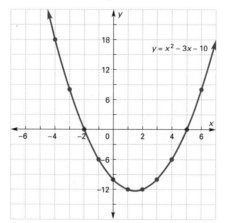

x	y
-4	18
-3	8
-2	0
-1	-6
0	-10
1	-12
2	-12
3	-10
4	-6
5	0
6	8

The roots of the quadratic equation $y = x^2 - 3x - 10$ are the values of x when the value of y is zero. Thus, you find the roots when you solve the quadratic equation

$$x^2 - 3x - 10 = 0$$

In Chapter 10, you learned how to rewrite a quadratic expression in factored form:

$$x^2 - 3x - 10 = (x - 5)(x + 2)$$

Thus, you can substitute the factored form of $x^2 - 3x - 10$ in the quadratic equation.

$$x^2 - 3x - 10 = 0$$

$$(x - 5)(x + 2) = 0$$

The substitution results in an equation that equates the product of two binomials to zero. If the product of two numbers is zero, what do you know about the numbers?

Zero Product Property

Let a and b be any real numbers.

If $ab = 0$, then $a = 0$ or $b = 0$.

To solve $(x - 5)(x + 2) = 0$, use the Zero Product Property. Set each factor equal to zero and solve the resulting linear equations.

$$\text{If } (x - 5)(x + 2) = 0, \text{ then}$$

$$x - 5 = 0 \quad \text{or} \quad x + 2 = 0$$

$$x = 5 \quad \text{or} \quad x = -2$$

You have solved $x^2 - 3x - 10 = 0$ by factoring. The solutions are 5 and -2. The solutions are the same x-intercepts shown on the graph.

EXAMPLE 1 The Roots of a Quadratic Equation

Find the roots of $y = 8x^2 - 30x + 25$.

SOLUTION

To find the roots of a quadratic equation, substitute zero for y and solve the resulting equation.

$8x^2 - 30x + 25 = 0$	Set original equation equal to zero
$(2x - 5)(4x - 5) = 0$	Factor the trinomial
$2x - 5 = 0 \quad \text{or} \quad 4x - 5 = 0$	Zero Product Property
$2x = 5 \quad \text{or} \quad 4x = 5$	Solve each linear equation
$x = \dfrac{5}{2} \quad \text{or} \quad x = \dfrac{5}{4}$	The solutions $\dfrac{5}{2}$ and $\dfrac{5}{4}$ are the roots

Substitute the solutions into the original equation $y = 8x^2 - 30 + 25$. Did you get $y = 0$?

Ongoing Assessment

Find the roots of $y = 3x^2 + 7x + 2$. Check the solution by substitution and by graphing.

EXAMPLE 2 Solving a Quadratic Equation

Solve $x^2 = 5x - 6$.

SOLUTION

Use the Addition Property of Equality to rewrite the equation in the form $x^2 + bx + c = 0$. Then factor and use the Zero Product Property.

$x^2 = 5x - 6$ Given

$x^2 - 5x + 6 = 0$ Addition Property of Equality

$(x - 3)(x - 2) = 0$ Factor the trinomial

$x - 3 = 0$ or $x - 2 = 0$ Zero Product Property

$x = 3$ or $x = 2$ Addition Property of Equality

The solutions are 3 and 2. Check the solutions in the original equation.

Critical Thinking   Explain how to solve $x^2 - 5x + 6 = 0$ by graphing $y = x^2$ and $y = 5x - 6$ on a graphics calculator.

EXAMPLE 3   Solving a Practical Problem

The Chambers family has a garden that measures 15 feet by 30 feet. They plan to increase the area of the garden by 250 square feet. However, they must increase the length and the width by the same number of feet. What are the dimensions of the garden after the increase?

SOLUTION

Draw a picture of the original garden.

Draw a picture of the new garden.

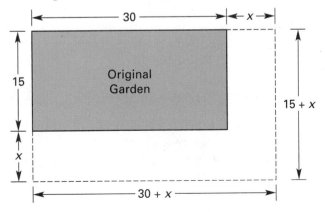

The area of the new garden is 450 + 250, or 700 square feet. Thus, the area of the new garden is modeled by this equation.

$$(30 + x)(15 + x) = 700$$

However, you cannot use the Zero Product Property with the equation written in this form. To use the Zero Product Property, first multiply the binomials. Then combine the constant terms and rewrite the equation with zero on the right side of the equal sign.

$(30 + x)(15 + x) = 700$	Given
$450 + 45x + x^2 = 700$	Multiply the binomials
$x^2 + 45x - 250 = 0$	Set trinomial equal to zero
$(x - 5)(x + 50) = 0$	Factor the trinomial
$x = 5 \text{ or } x = -50$	Zero Product Property

There are two possible solutions. The Chambers family can increase the garden by 5 feet on each side, or they can decrease it by 50 feet on a side. You can see from the drawing that a decrease of 50 feet is impossible. The solution to an equation does not always make sense in a real world problem. Always check the domain and the range before you give the answer. Does a solution of 5 make the equation true? Look at the drawing. Does the solution make sense?

LESSON ASSESSMENT

Think and Discuss

1 Explain how to find the roots of a quadratic equation by factoring.

2 How is the Zero Product Property used to solve a quadratic equation?

Practice and Problem Solving

Solve each quadratic equation by factoring. Check your answer.

3. $x^2 + 5x + 6 = 0$ **4.** $x^2 + 5x + 4 = 0$ **5.** $x^2 + 2x = 3$

6. $x^2 - 2x = 3$ **7.** $x^2 + 3x + 2 = 0$ **8.** $x^2 + 3x = 10$

9. $2x^2 + 5x = 12$ **10.** $3x^2 = 8x + 3$ **11.** $x^2 = 16 - 6x$

12. $4x^2 + 2x = 2$ **13.** $6x^2 + x - 12 = 0$ **14.** $4x^2 = 4x - 1$

Write and solve an equation for each situation.

15. The number of square meters in the area of a square deck is 12 greater than the number of meters around the deck's perimeter. What is the length of each side of the deck?

16. Shemekia is cutting material for a triangular sail. The height of the sail must be 2 feet longer than the base. The area of the sail is 40 square feet. What dimensions should Shemekia cut for the sail?

Mixed Review

The amount of matter in a body is called its mass. Scientific notation is used to approximate very large and very small masses. For example,

the mass of a human cell is about 10^{-18} kg
the mass of a paper clip is about 10^{-3} kg
the mass of the moon is about 6×10^{22} kg

17. How many times greater is the mass of the moon than the mass of a paper clip? Write a ratio in simplified form comparing the two masses.

18. How many times less is the mass of a human cell than the mass of a paper clip? Write a ratio in simplified form comparing the two masses.

The length of a rectangle is one foot longer than its width. The perimeter is 180 inches.

19. Use one unknown to write an equation for the perimeter.

20. Solve the equation in Exercise 19.

21. Use two unknowns to write a system of equations for the perimeter.

22. Solve the system in Exercise 21 by substitution.

23. How are the solutions in Exercise 20 and 22 similar?

24. Which method of solution seems easier? Explain your answer.

The function $f(t) = -16t^2 + 400$ represents the height of a freely falling ballast bag that starts from rest on a balloon 400 feet above the ground.

25. Find $f(t)$ when t is 5.

26. What does this solution mean with regard to the height of the object?

LESSON 11.4 COMPLETING THE SQUARE

In Lesson 10.7, you used geometric models to factor quadratics such as $x^2 + 4x + 4$.

In this lesson you will learn a technique that can be used to solve any quadratic equation. This technique is called **completing the square**.

Start with the model for $x^2 + 6x$. How many 1-blocks must you add to one x^2-block and six x-blocks to build a square?

x^2 $6x$

ACTIVITY 1   Completing a Square

Draw an area diagram for each step.

1 Start with one x^2-block and six x-blocks.

2 Divide the six x-blocks into two equal sets.

3 How many blocks are in each set?

4 Square the number found in Step 3. Add this number of 1-blocks to the area diagram.

5 Form a square using all the blocks. What are the dimensions of the square?

6 Complete the Activity again. This time start with

    a. $x^2 + 8x$ **b.** $x^2 + 10x$

In Activity 1, you used a geometric method to complete a square. There is also an algebraic method for completing the square of a quadratic equation in the form $y = x^2 + bx$. First, divide the coefficient of the x-term by 2. The result is $\frac{b}{2}$. Next, square the result of the division. The product is $\left(\frac{b}{2}\right)^2$. Finally, add $\left(\frac{b}{2}\right)^2$ to both sides of the equation.

$$y + \left(\frac{b}{2}\right)^2 = x^2 + bx + \left(\frac{b}{2}\right)^2$$

The right side of the equation is now a perfect square trinomial.

$$x^2 + bx + \left(\frac{b}{2}\right)^2 = \left(x + \frac{b}{2}\right)^2$$

Ongoing Assessment

Complete the square for $x^2 + 100x$.

ACTIVITY 2   The Vertex of a Parabola

1 Use a table or graphics calculator to graph $y = x^2 - 6x$. What are the coordinates of the vertex?

2 What number should be added to make a perfect square trinomial from $x^2 - 6x$? Add this number to both sides of the equation. What property lets you do this without changing the equality?

3 Factor the perfect square trinomial.

4 Solve the equation for y. How can you use this equation to write the vertex of the parabola?

In Activity 2, you completed the square to write the equation

$$y = x^2 - 6x \ \text{ as } \ y = (x - 3)^2 - 9$$

which is in the form $y = (x - h)^2 + k$. Using your knowledge of transformations, you can see that the vertex of a parabola is the point (h, k). Thus, the vertex is $(3, -9)$.

When a quadratic equation is written in the form

$$y = a(x - h)^2 + k$$

the values of a, h, and k provide useful information for sketching the resulting parabola.

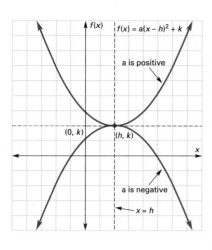

a is positive

(0, k)

(h, k)

a is negative

x = h

The Parabola Form of a Quadratic Function

The parabola form of a quadratic equation is

$$f(x) = a(x - h)^2 + k$$

The graph of f is a parabola with vertex (h, k). When a is positive, the *minimum* value of f is k. When a is negative, the *maximum* value of f is k. The axis of symmetry is the line $x = h$.

EXAMPLE 1 Completing the Square to Find the Vertex

Find the vertex of $y = x^2 + x - 12$ by completing the square.

SOLUTION

First, isolate the $x^2 + bx$ terms. Then, complete the square. Finally, write the equation in parabola form.

$y = x^2 + x - 12$	Given
$y + 12 = x^2 + x$	Addition Property of Equality
$y + 12 + \left(\dfrac{1}{2}\right)^2 = x^2 + x + \left(\dfrac{1}{2}\right)^2$	Complete the square
$y + 12.25 = (x + 0.5)^2$	Factor a perfect square
$y = (x + 0.5)^2 - 12.25$	Subtraction Property of Equality

Compare the final equation to the parabola form of a quadratic. In this form, h is -0.5 and k is -12.25. The vertex is $(-0.5, -12.25)$.

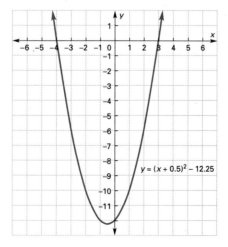

$y = (x + 0.5)^2 - 12.25$

Find the vertex of $y = x^2 - x - 12$ by completing the square.

EXAMPLE 2 Finding the Minimum of a Quadratic

Find the minimum of $f(x) = x^2 + 3x$.

SOLUTION

Solve the equation $y = x^2 + 3x$ by completing the square.

$$y = x^2 + 3x \qquad\qquad \text{Given}$$

$$y + \left(\frac{3}{2}\right)^2 = x^2 + 3x + \left(\frac{3}{2}\right)^2 \qquad \text{Complete the square}$$

$$y + \frac{9}{4} = \left(x + \frac{3}{2}\right)^2 \qquad \text{Factor a perfect square}$$

$$y = \left(x + \frac{3}{2}\right)^2 - \frac{9}{4} \qquad \text{Subtraction Property of Equality}$$

The quadratic is now written in parabola form. The vertex (h, k) is at $\left(-\frac{3}{2}, -\frac{9}{4}\right)$. Since a is positive, the vertex is the minimum value for the function. Thus, the minimum of the function occurs at $f\left(-\frac{3}{2}\right)$ or $-\frac{9}{4}$.

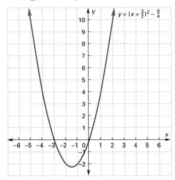

EXAMPLE 3 Finding the Maximum of a Quadratic

Find the maximum of $f(x) = -x^2 + 6x - 4$.

SOLUTION

Solve the equation $y = -x^2 + 6x - 4$ by completing the square.

$$y = -x^2 + 6x - 4 \qquad \text{Given}$$

$$y + 4 = -(x^2 - 6x) \qquad \text{Addition Property of Equality and Distributive Property}$$

$$y + 4 - 9 = -(x^2 - 6x + 9) \qquad \text{Complete the square and Addition Property of Equality}$$

$$y - 5 = -(x - 3)^2 \qquad \text{Factor perfect square trinomial}$$

$$y = -(x - 3)^2 + 5 \qquad \text{Addition Property of Equality}$$

Compare this equation to the parabola form of the quadratic equation. The vertex (h, k) is at $(3, 5)$. Since a is negative, the vertex is the maximum point on the graph. Thus, the maximum of the function occurs for $f(3)$ or 5.

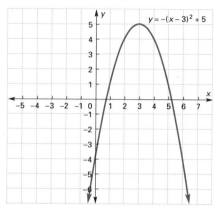

$y = -(x - 3)^2 + 5$

Ongoing Assessment

Find the minimum or maximum of $f(x) = -x^2 - 8x + 1$.

WORKPLACE COMMUNICATION

Divide both sides of the equation for total area by 4π. Assume a value for the height, such as $h = 4$ meters. Then, complete the square of the quadratic equation. Where is the vertex? Can you minimize A, as stated by Mr Bakorsi? How will you respond to Mr. Bakorsi? What other factors might influence your choice for r and h?

 Modern Workplace, Inc.
Interoffice Memo

To: Ruby Mahan, Hazardous Material Handling Team Leader
From: Jack Bakorsi, Director of Plans and Programs
Subject: Optimum Design for Liquid Nitrogen Storage Tanks

I understand that you are planning to redesign the current liquid nitrogen storage tanks. Before you begin construction of the replacements, please design the most cost effective, efficient tank. I think this is the tank with minimum surface area.

To fit the constraints of the storage facility, the shape of the tanks must be right-circular cylinders with hemispherical end-caps. You cannot control the shape, but you can control the height and the radius and, therefore, the total surface area of the tanks.

Surface Area of
Spherical End-Caps $= 4\pi r^2$

Surface Area of
Cylinder $= 2\pi rh$

The surface area of the tank is the sum of the areas of the end-caps and the cylinder:

$$A = 4\pi r^2 + 2\pi rh$$

Since this is a quadratic equation, you should be able to minimize the surface area A. This will minimize the cost of the insulation surrounding the tank and will minimize the heat loss from the liquid nitrogen. This should greatly reduce the number of times we have to refill the tank each year.

In the long run, we can save the company a lot of money by using the optimum values of r and h for the new liquid nitrogen storage tanks. Please let me know what values you plan to use.

LESSON ASSESSMENT

Think and Discuss

Pick a value for b and substitute it into the equation $y = x^2 + bx$. Use this function to answer 1-2.

1 Explain how to use a geometric model to complete the square of your function.

2 Explain how to use the method of completing the square to find the vertex of the graph of your function.

3 Explain how to use the method of completing the square to find the vertex of the parabola formed by the graph of $y = x^2 + 5x + 6$.

Practice and Problem Solving

For each quadratic equation,

a. write the equation in parabola form $y = a(x - h)^2 + k$.

b. find the minimum or maximum value of the function.

c. find the coordinates of the vertex.

d. write the equation for the line of symmetry.

4. $y = x^2 + 4x$

5. $y = -x^2 + 8x$

6. $y = x^2 - 10x$

7. $y = -x^2 + 3x + 2$

8. $y = x^2 + 6x + 8$

9. $y = -x^2 - 7x + 10$

10. $y = x^2 - 3x - 4$

11. $y = -x^2 - 4x - 12$

12. $y = x^2 - 5x + 6$

13. Suppose you earn $160 in simple interest during one year on a $2000 investment. What is the rate of interest?

14. The diameter of a baseball is about 7.6 centimeters. The diameter of a basketball is about 24 centimeters. Find the difference between the volume of a baseball and a basketball.

Write and solve a system of linear equations to model each situation.

15. The Right Company manufactures hammers for $4 each and wrenches for $2 each. On Tuesday the Right Company manufactured a total of 960 hammers and wrenches at a cost of $2880. How many hammers did the Right Company manufacture on Tuesday?

16. The Great Coffee Shop's bestselling coffees are hazelnut and Brazilian. A package containing 2 pounds of hazelnut and 3 pounds of Brazilian costs $34.50. A package containing 3 pounds of hazelnut and 2 pounds of Brazilian costs $35.50. How much does one pound of each coffee cost?

LESSON 11.5 SOLVING QUADRATIC EQUATIONS

In previous lessons, you used graphing and factoring to find the zeros of a quadratic function such as $f(x) = x^2 - 2x - 3$. You can also find the zeros by completing the square.

EXAMPLE 1 Finding the Zeros of a Function

Find the zeros of $f(x) = x^2 - 2x - 3$ by completing the square.

SOLUTION

The zeros of the function occur when $f(x) = 0$.

$x^2 - 2x - 3 = 0$	Given
$x^2 - 2x = 3$	Addition Property of Equality
$x^2 - 2x + 1 = 3 + 1$	Completing the Square and Addition Property of Equality
$(x - 1)^2 = 4$	Factoring
$x - 1 = 2$ or $x - 1 = -2$	If $x^2 = c, x = \pm \sqrt{c}$
$x = 3$ or $x = -1$	Addition Property of Equality

The zeros are 3 and -1.

Check:

$$f(3) = (3)^2 - 2(3) - 3 = 9 - 6 - 3 = 0$$

$$f(-1) = (-1)^2 - 2(-1) - 3 = 1 + 2 - 3 = 0$$

Consider the function $f(x) = x^2 - 5x - 14$. How can you find the value or values of x such that $f(x) = -20$?

You can rewrite the equation $x^2 - 5x - 14 = -20$ as $x^2 - 5x = -6$. Then solve $x^2 - 5x = -6$ by any one of the methods you have learned in this chapter.

1. Let $f(x) = -20$. Rewrite $f(x) = x^2 - 5x - 14$ as the quadratic equation $x^2 - 5x - 14 = -20$.

2. What property allows you to rewrite $x^2 - 5x - 14 = -20$ as $x^2 - 5x = -6$?

3. Solve $x^2 - 5x = -6$ by completing the square.

4. Solve $x^2 - 5x = -6$ by factoring. (*Hint:* add 6 to both sides.)

5. Solve $x^2 - 5x = -6$ by graphing. (Use a graphics calculator if available.)

6. Which method do you prefer? Explain why.

EXAMPLE 2   Finding the Dimensions of a Rectangle

Taj wants to construct a rectangular cabin floor that has a perimeter of 54 feet and an area of 180 square feet. What dimensions should Taj use to construct the cabin?

SOLUTION

Let x represent the length of the cabin floor. Let y represent the width of the cabin floor.

$$\boxed{\text{cabin floor}} \quad y$$
$$x$$

The perimeter (54) of the cabin floor is $p = 2x + 2y$. Thus,

$$2x + 2y = 54 \quad \text{or} \quad x = 27 - y$$

The area (180) of the cabin floor is $A = xy$. Thus,

$$A = (27 - y)y \quad \text{or} \quad 180 = 27y - y^2$$

The solution will be the roots of this quadratic equation.

$$y^2 - 27y + 180 = 0$$

You can use any of the three methods to solve this quadratic equation. Since the roots are 15 and 12, the dimensions of the cabin floor should be 15 feet by 12 feet.

In an auditorium, the number of chairs in each row is 8 fewer than the number of rows. The auditorium holds 609 people. How many rows are in the auditorium?

Sometimes the solution to a quadratic equation is an irrational number, and you must approximate the roots.

EXAMPLE 3 Irrational Roots

Solve the quadratic $x^2 - 4x = 2$.

SOLUTION

$$x^2 - 4x = 2 \qquad \text{Given}$$

$$x^2 - 4x + 4 = 2 + 4 \qquad \text{Completing the Square and}$$
$$\text{Addition Property of Equality}$$

$$(x - 2)^2 = 6 \qquad \text{Factoring}$$

$$x - 2 = \sqrt{6} \ \text{ or } \ x - 2 = -\sqrt{6} \qquad \text{If } x^2 = c, x = \pm\sqrt{c}$$

$$x = 2 + \sqrt{6} \ \text{ or } \ x = 2 - \sqrt{6} \qquad \text{Addition Property of Equality}$$

The roots are irrational because they cannot be written as the ratio of two integers. Use your calculator to find that the roots are approximately 4.45 and -0.45.

LESSON ASSESSMENT

Think and Discuss

Explain how to solve the quadratic equation $y = x^2 - 6x + 8$ by each of the following methods.

1 Graphing.

2 Factoring.

3 Completing the square.

Practice and Problem Solving

Solve each equation by completing the square. Round any irrational roots to the nearest hundredth.

4. $x^2 - 2x - 15 = 0$ **5.** $x^2 = 4x + 12$ **6.** $x^2 = 6x - 5$

7. $x^2 - 4x = 7$ **8.** $x^2 = 6x - 9$ **9.** $x^2 = 7 - 8x$

10. $x^2 + 12x = 0$ **11.** $x^2 = x + 3$ **12.** $x^2 + 3x = 3$

13. $x^2 - 3 = 5x$ **14.** $x^2 = x$ **15.** $x^2 = 1 - 4x$

Let $f(x) = x^2 - 8x + 10$. Find the value of x when $f(x)$ is the given value.

16. $f(x) = -5$ **17.** $f(x) = 10$ **18.** $f(x) = -2$

Solve using quadratic equations.

19. The length of a playing field is 2 meters longer than its width. If the area of the playing field is 1200 square meters, find the dimensions of the playing field.

20. The sum of two whole numbers is 21. Their product is 108. Find the numbers.

Mixed Review

The formula $P = R - B - C$ is used by insurance companies to find their profit (P). R is the revenue received as premiums. B is the benefits paid out. C is the overhead cost.

21. The Universal Insurance Company receives $400 in premiums for a certain life insurance policy. The overhead cost for the policy is $25 per policy. During one year, the company sold 20,000 policies and paid out $6.45 million in benefits. What is the profit for this policy?

22. Solve the formula for the overhead cost.

A dart is tossed and hits the dartboard in random fashion.

23. What is the probability the dart lands in section A?

24. What is the probability the dart lands in section B?

25. What is the probability the dart lands in section C?

26. What is the probability the dart does *not* land in section D?

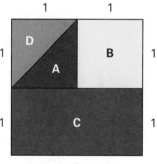

LESSON 11.6 THE QUADRATIC FORMULA

When a quadratic equation is written in the form

$$y = ax^2 + bx + c$$

you can find the roots by graphing, factoring, or completing the square. There is one more method you can use to find the roots. This method uses the **quadratic formula**. This formula can be derived by solving $ax^2 + bx + c = 0$ for x.

ACTIVITY 1   Deriving the Quadratic Formula

Give the reason for each step.

$$ax^2 + bx + c = 0$$

$$ax^2 + bx = -c$$

$$x^2 + \frac{b}{a}x = -\frac{c}{a}$$

$$x^2 + \frac{b}{a}x + \frac{b^2}{4a^2} = -\frac{c}{a} + \frac{b^2}{4a^2}$$

$$\left(x + \frac{b}{2a}\right)^2 = -\frac{c}{a} + \frac{b^2}{4a^2}$$

$$\left(x + \frac{b}{2a}\right)^2 = \frac{4ac}{4a^2} + \frac{b^2}{4a^2}$$

$$\left(x + \frac{b}{2a}\right)^2 = \frac{b^2 - 4ac}{4a^2}$$

$$x + \frac{b}{2a} = \pm\sqrt{\frac{b^2 - 4ac}{4a^2}}$$

$$x = \frac{\pm\sqrt{b^2 - 4ac}}{2a} - \frac{b}{2a}$$

$$x = \frac{-b \pm \sqrt{b^2 - 4ac}}{2a}$$

You can use the quadratic formula to find the roots of any quadratic equation written in the form $ax^2 + bx + c = 0$.

The Quadratic Formula

Let a, b, and c be real numbers with $a \neq 0$.

If a quadratic equation is written in the form

$ax^2 + bx + c = 0$, then the roots of the equation are

$$x = \frac{-b \pm \sqrt{b^2 - 4ac}}{2a} \qquad \text{where } a \neq 0$$

EXAMPLE 1 Applying the Quadratic Formula

Find the x-intercepts of $y = 8x^2 - 30x + 25$.

SOLUTION

To find the x-intercepts, first substitute zero for y. Then use the quadratic formula to solve $8x^2 - 30x + 25 = 0$. In the formula, a is 8, b is -30, and c is 25.

$$y = 8x^2 - 30x + 25$$

$$y = ax^2 - bx \ + c$$

$$x = \frac{-b \pm \sqrt{b^2 - 4ac}}{2a}$$

$$x = \frac{-(-30) \pm \sqrt{(-30)^2 - (4)(8)(25)}}{(2)(8)}$$

$$x = \frac{30 \pm \sqrt{900 - 800}}{16}$$

$$x = \frac{30 \pm \sqrt{100}}{16}$$

$$x = \frac{30 \pm 10}{16}$$

$x = 2.5$ (when $+ 10$ is used)

$x = 1.25$ (when $- 10$ is used)

The table shows the pattern that occurs when a number of people talk once to each other on the telephone.

Number of People	Number of Phone Calls	Geometric Representation of Phone Calls
1	0	•
2	1	
3	3	
4	6	
5	10	
etc.	etc.	etc.

You can use a formula to express the relationship between the number of phone calls completed and the number of people talking to each other.

Let c represent the number of phone calls.
Let p represent the number of people making calls.

The following equation models the relationship in the table:

$$c = \frac{p(p - 1)}{2}$$

1 Let c represent 45 phone calls. Substitute 45 for c in the equation

$$c = \frac{p(p - 1)}{}$$

2 Show how to write the new equation as $p^2 - p - 90 = 0$.

3 Solve the quadratic equation using the quadratic formula.

4 Check each root in the original equation.

5 If 45 phone calls are completed, how many people talked to each other?

WARNING

This home is protected by SPARTAN SECURITY SYSTEMS

EXAMPLE 2 The Profit Function

Spartan Security installs security systems. To track monthly revenues for the number of systems installed (x), they use the function

$$R(x) = 10x^2 + 10x + 100$$

Spartan Security projects their costs using the function

$$C(x) = 8x^2 + 50x + 350$$

Find a function to project monthly profit and use it to find the break-even point.

SOLUTION

To find the profit function, $P(x)$, subtract the cost function from the revenue function.

$$P(x) = R(x) - C(x)$$

$$P(x) = (10x^2 + 10x + 100) - (8x^2 + 50x + 350)$$

$$= 2x^2 - 40x - 250$$

The break-even point occurs when revenues equal cost—in other words, when $P(x) = 0$.

Thus, the roots of $2x^2 - 40x - 250 = 0$ will give the break-even point. If you use the quadratic formula with $a = 2, b = -40$, and $c = -250$, the result is $x = 25$ or $x = -5$. Since there cannot be a negative number of security systems, the only root of the quadratic that makes sense is 25. The company will break even after it installs 25 systems.

The quadratic formula gives more than just the roots of quadratic equations. The portion of the quadratic formula under the radical, $b^2 - 4ac$, is called the **discriminant**. In Example 1, the discriminant is 100. When the discriminant is greater than zero, the equation has two unequal roots.

$$x = \frac{-b \pm \sqrt{b^2 - 4ac}}{2a}$$

Critical Thinking Why does a positive discriminant mean the graph of a quadratic equation intersects the x-axis twice?

Consider the quadratic $y = x^2 - 8x + 16$.

When you use the quadratic formula with $a = 1$, $b = -8$, and $c = 16$, the following result is obtained.

$$x = \frac{-b \pm \sqrt{b^2 - 4ac}}{2a}$$

$$x = \frac{-(-8) \pm \sqrt{(-8)^2 - (4)(1)(16)}}{(2)(1)}$$

$$x = \frac{8 \pm \sqrt{64 - 64}}{2} = \frac{8 \pm 0}{2}$$

$$x = 4$$

The graph of $y = x^2 - 8x + 16$ intersects the x-axis at exactly one point: $(4, 0)$. When the discriminant is equal to zero, the equation has exactly one root.

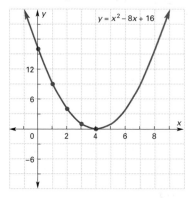

x	y
0	16
1	9
2	4
3	1
4	0

Now examine the graph of the equation $y = 3x^2 + 4x + 5$.

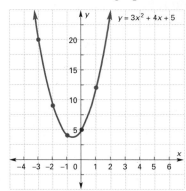

x	y
−3	20
−2	9
−1	4
0	5
1	12
2	25

Notice that the graph does not intersect the x-axis. There are no real numbers that make the equation true when y is equal to zero. Also notice that the discriminant is $(4)^2 - (4 \cdot 3 \cdot 5)$, or -44. If the discriminant is less than zero, there are no real roots.

EXAMPLE 3 Using the Discriminant

Find the discriminant of the equation $y = 2x^2 + 5x + 15$. Use the discriminant to describe the roots.

SOLUTION

For the discriminant, a is 2, b is 5, and c is 15. The discriminant $b^2 - 4ac$ is $25 - 120$, or -95. Since the discriminant is less than zero, the equation $2x^2 + 5x + 15 = 0$ has no real roots. Thus, the expression $2x^2 + 5x + 15$ is prime.

CULTURAL CONNECTION

The Great Pyramid at Giza utilizes a special ratio between the altitude of a triangular face and one-half the length of its base. This ratio is known as the *golden ratio* and has been used by artists and architects to create many works of art over the centuries.

To find the golden ratio, start with segment AC and point B that divides segment AC in the following way:

$$\frac{a}{x} = \frac{x}{x+a}$$

The ratio of the shorter segment AB to the longer segment BC is equal to the ratio of the longer segment BC to the total length AC.

1. Let $a = 1$. Solve the quadratic equation that results when you cross-multiply. The positive root is the golden ratio.

Research the relationship between the golden ratio and

2. the Parthenon.

3. a chambered nautilus.

4. the Fibinocci sequence.

5. the Pentagon in Washington, D.C.

LESSON ASSESSMENT

Think and Discuss

1 How is the quadratic formula used to find the roots of a quadratic equation?

2 How does the discriminant describe the roots of a quadratic function?

Identify *a*, *b*, and *c* for each quadratic equation. Find the value of the discriminant.

3. $3x^2 = 5x - 1$ **4.** $4x^2 - 16 = 0$ **5.** $-x^2 = 2x - 1$

Use the quadratic formula to solve each equation. Use your calculator to approximate roots to the nearest hundredth when necessary. Indicate if there are no real roots.

6. $x^2 - 2x - 3 = 0$ **7.** $x^2 - 4x = 21$ **8.** $3x^2 - 2x = 1$

9. $x^2 = x - 6$ **10.** $-x^2 + 6x = 9$ **11.** $8x^2 = 2 - 7x$

12. $2x^2 = 2x + 5$ **13.** $3x^2 = x - 1$ **14.** $0.5x^2 = 0.2x + 0.1$

Solve using a quadratic equation.

15. The length of a rectangular garden is 5 meters longer than its width. If the area of the garden is 84 square meters, find the dimensions of the garden.

16. A rectangular garden has an area of 76 square feet. If the perimeter of the garden is 80 feet, what are the dimensions of the garden?

17. A piece of copper tubing 60 inches long is cut into two unequal pieces. Each piece is formed into a square frame. If the total area of the two frames is 117 square inches, what is the length of each piece of tubing?

Marguerite Gomez owns a company that makes sensors for air bags. She uses the function $P(x) = x^2 - 280x - 14{,}625$ to estimate her yearly profit. $P(x)$ represents the profit, and x represents the number of sensors she sells.

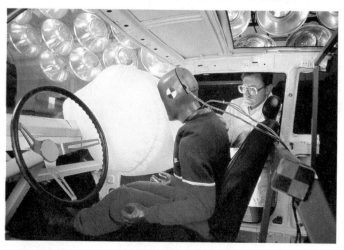

18. How many sensors must Ms. Gomez sell to break even?

19. How many sensors must Ms. Gomez sell to make a profit of $1125?

20. Ms. Gomez's company has the capacity to produce 500 sensors per year. If the company operates at full capacity, what is the expected profit?

The distance (d) in feet an object falls in t seconds is
$d = 16t^2 + 5t$.

21. About how long will it take for an object to fall 1000 feet?

22. About how long will it take for an object to fall 4000 feet?

Mixed Review

The Lakeside Running Club recorded the following distances in miles jogged by its members in one week:

14 18 18 16 16 23 21 15 20 25
15 16 22 24 21 16 25 23 17 15

23. Find the range of the data.

24. Find the mode of the data.

25. Find the median of the data.

26. Make a frequency distribution table using intervals of three units.

27. Find the mean of the data.

28. Use the frequency distribution table to draw a histogram.

MATH LAB

Activity 1: Measuring a Parabolic Reflector

Equipment   Flashlight reflector, preferably from a 6 volt
flashlight
Vernier caliper
Ruler with centimeter scale
Masking tape

Problem Statement

Reflectors of light and radio waves often have the shape of a parabola. Parabolic reflectors are useful because incoming radio waves reflect to a common point, called the *focus*. In the opposite fashion, if a bright light source is located right at the focus, its light rays will reflect away from the reflector as parallel rays. The illustrations below show a cross-sectional view of a reflector and a collector.

Reflector

Collector

In the case of the reflector, a light bulb is positioned at a point called the *focal point, f*. Light from the focal point is reflected out in rays parallel to the axis of the parabola. When used as a collector, the parabolic mirror takes incoming light rays that are parallel to the axis and reflects them to the focal point, *f*.

You will measure the coordinates of a flashlight reflector to determine if its shape is really parabolic. You will locate the focus of the reflector and determine if the bulb's filament is at the focal point of the reflector.

Procedure

a Disassemble the flashlight so that the reflector is exposed. (Do not remove the bulb.)

b Lay the ruler flat across the top lip of the reflector so that the edge of the centimeter scale is directly above the center of the bulb. Use masking tape to hold the ruler in place

Positioning ruler
atop reflector

Measuring depth on left side
with vernier caliper

c Practice measuring the depth of the reflector as shown in the illustration. Be careful not to scratch the reflector surface with the tip of the caliper probe. Use the following technique to take at least three measurements on each side of the bulb:

1. Open the caliper to 1.0 cm.

2. Carefully place the probe into the reflector while resting the edge of the caliper on the ruler (see illustration).

3. Slide the caliper along the ruler until the tip just touches the reflector. If the probe has an uneven tip, be sure the tip of the longest edge touches the reflector surface.

4. With the caliper probe held in a vertical position (that is, perpendicular to the ruler's edge), read the measurement on the ruler (to the nearest 0.1 cm) where the probe contacts the ruler.

5. Record the measurement from the ruler as the *x*-value and the preset vernier caliper reading as the *y*-value.

Ruler Position x (cm)	Vernier Caliper Setting y (cm)

6. With the same setting, move the vernier to the opposite side of the bulb and obtain a measurement from the ruler. Record these x- and y-values in the Data Table. (You should have two x-measurements for the same y-value.)

7. Repeat Steps 1–6, each time increasing the opening of the caliper about 1.0 cm. You should have at least three measurements on each side of the bulb.

d Before removing the ruler, use the vernier caliper probe to estimate the location of the filament. Estimate the x-value by looking at the ruler position directly above the bulb. Find the y-value by positioning the vernier caliper probe beside the bulb, with the tip at the same height as the filament. Record both the x-value (the ruler position) and the y-value (the vernier caliper reading).

e Remove the masking tape and the ruler and reassemble the flashlight.

f Graph the x- and y-values you recorded for the reflector.

g Plot the point you measured for the bulb filament. This point should be close to the focus of the parabola. Complete the following steps to plot the first point on a line called the *directrix*.

    1. Draw a line from the focus to a point you plotted in **f**.

    2. Measure the length of this line in centimeters.

    3. From the plotted point, draw a vertical line segment of the same length. Darken the endpoint at the bottom of this line segment.

h Repeat Steps 1–3 of **g** for each of the points you plotted. Draw a horizontal line that best fits the dark points. This line will estimate the position of the directrix. Your graph should look similar to the one shown here.

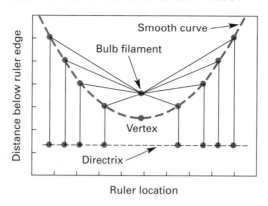

i Draw a final point halfway between the focus and the directrix. This is the vertex of the parabola.

j Draw a smooth curve through your data points and through the vertex. Does your curve resemble a parabola?

k Answer the following questions.

1. If the bulb filament is really *not* at the focus but a little high (that is, too far from the vertex), will the dark points form a straight line? Explain.

2. What do you think would be the effect on the flashlight beam if the bulb were too high?

3. Suppose you have a satellite dish antenna. You know that it is shaped like a parabola. Describe how you would locate the focus so that you could accurately position the receiver element.

Equipment   Heavy construction paper
Ruler

Problem Statement

You need to construct a rain gutter from a rectangular piece of sheet metal (simulated in this Activity by a piece of paper) that is 12 inches wide.

To form the gutter, you will make a 90° bend on each of the two sides of the sheet metal to form a U-shaped gutter. For good drainage, the gutter must have a cross-sectional area of at least 15 square inches. You will determine the length of the gutter sides to meet this area requirement.

Procedure

a Use your ruler to place marks at 1-inch intervals along both 12-inch sides of your construction paper. Draw a line between each pair of marks.

b Make a 90° fold along the outer pair of lines to "make a gutter" with 1-inch sides. Make a sketch of the cross section of the gutter. Record the height (h) of the sides, the width (w) of the base, and the cross-sectional area. Include proper units in your data.

Side (*h*)	Width (*w*)	Area (*A*)

c Smooth out the folds made in the previous step. Repeat Step **b** for each of the other pairs of lines to make "gutters" with sides of 2 inches, 3 inches, and so on.

d Examine your data. Which of the folds would make a gutter that has at least 15 square inches of cross-sectional area?

e Draw a graph of your measurements. Let the horizontal axis represent the size of the fold. Let the vertical axis represent the cross-sectional area.

f Identify the portion of the graph that will produce a gutter with a cross-sectional area of at least 15 square inches.

g Write an equation for the width (*w*) of the gutter in terms of the height (*h*). Write a second equation that expresses the gutter's cross-sectional area (*A*) in terms of *w* and *h*. Substitute the expression for the width from the first equation into the second equation. Rewrite the resulting equation in standard form for a quadratic equation.

h Find the values of *h* that will yield an area *A* = 15. You can use the quadratic formula or the method of completing the square to determine the values of *h* that will make the equation true when you substitute *A* = 15 in your quadratic equation.

i What value of *h* makes the cross-sectional area a maximum? (*Hint:* Use your graph to help answer this question.)

j Would your graph be any different if you had chosen the width (*w*) to be the horizontal axis instead of the height (*h*)? Explain.

Equipment   Calculator
Computer with BASIC installed
Tape measure

Problem Statement

In this Activity, you will find the roots of a quadratic equation using a computer program. The quadratic equation will model the area of your teacher's desktop. The solution or roots to this equation will provide information for equally expanding the dimensions of the desktop to double its area.

Procedure

a Measure the length and the width of the top of your teacher's desk. Record the measurements in the Data Table below.

Object	Length (feet or meters)	Width (feet or meters)	Area (Length × Width) (square feet or square meters)
Desktop			

b Calculate the area of the desktop. Write this value in the Data Table.

c Write an expression for increasing the length of the desktop by an amount x. Write a similar expression for increasing the width of the desktop by the same amount x.

d Write an equation for the expanded area of the desktop using the expressions from Step **c**. Let the variable A represent the expanded area.

e Double the value of the area of the desktop calculated in Step **b**. Form a new equation by replacing A in the equation from Step **d**. with the value of the doubled area. Write the quadratic equation in the form of $ax^2 + bx + c = 0$.

f Write a BASIC computer program to find the roots of the equation in Step **e**. Use the program on the opposite page as a guideline.

```
10 REM PROGRAM TO PRINT VALUES OF F(X)
20 PRINT "(X, Y)"
30 PRINT
40 FOR X = d₁ TO d₂
```

(*Note:* d_1 and d_2 are selected upper and lower limits of the domain of the function that allow you to determine if there are any values of x which will yield a zero value for y.)

```
50 Y=(Enter your equation from Step e, with 0 replaced
```
by Y.)

```
60 PRINT"(";X;",,";Y;")"
70 NEXT X
80 END
```

g RUN the program.

h Between what two x-values do the y-values change from positive to negative? From negative to positive?

i Change line 40 of the computer program as follows:

```
40 FOR X = d₁ TO d₂ STEP .1
```

Note: d_1 and d_2 are the positive x-values found in Step **h** that caused the y-values to change from negative to positive.)

j RUN the new program.

k Repeat Steps **g** through **j** until you determine the positive x-value (to the nearest tenth) that results in a y-value of zero. (*Note:* the y-values will only change from negative to positive.)

l Explain the meaning of the x-value found in Step **k**.

m In Step **h**, you found two x-values between which the y-values changed from positive to negative. These x-values were not used to find the second root of the equation. Use either the quadratic formula or the BASIC program above to find the second root of the equation. Does this root have any physical meaning in the problem? Explain why or why not.

n Compare the area of the expanded desktop to that of the original desktop. Is the area doubled using the x-value found in Step **k**?

MATH APPLICATIONS

The applications that follow are like the ones you will encounter in many workplaces. Use the mathematics you have learned in this chapter to solve the problems.

Wherever possible, use your calculator to solve the problems that require numerical answers and use your graphics calculator to draw graphs.

Identify which of the following equations are quadratic functions. If the function is quadratic, sketch a graph or use your graphics calculator to find the maximum or minimum and the roots of the function.

1. $y = 12x - 7$

2. $y = x^2 - 8x + 15$

3. $y = x^3 + 5x + 6$

4. $y^2 = 2x^2$

5. $y = x^2 + 1$

6. $y = 2x^2 + 5x - 3$

The function $d = 9t$ represents the distance (d) traveled by a sprinter in time (t) when running at a constant speed of 9 meters per second. The function $d = 0.9t^2$ represents the distance traveled if the sprinter starts at rest and accelerates at a constant rate of 1.8 meters per second.

7 Describe the graph of each function.

8 Graph each function on the same coordinate axes.

9 At what point do the graphs intersect? What is the physical meaning of this point?

AGRICULTURE & AGRIBUSINESS

You want to construct a cylindrical tank for storing grain. You know that the formula for the surface area of a right-circular cylinder with no bottom is $A = \pi r^2 + 2\pi rh$, where A is the surface area, r is the radius of the cylinder, and h is the height of the cylinder.

10 The material for constructing the tank is a rectangular piece of sheet metal 12 feet wide and 40 feet long. What is the total area of available sheet metal?

11 Since the sheet metal is 12 feet wide, you must restrict yourself to a cylinder that is 12 feet high (that is, $h = 12$ feet). Equate the area of a cylinder to the area of sheet metal available. Write a quadratic equation in the form $ax^2 + bx + c = 0$, where x represents the unknown radius of the cylindrical tank.

12 Use the quadratic formula to determine the radius of a tank that will use all of the sheet metal.

13 What is "physically" wrong with the answer to Exercise 12? How might you overcome this problem in your equation? (*Hint:* Think of cutting a circular top from the rectangular sheet. Draw a diagram.)

A farmer normally plows a rectangular field 80 meters by 120 meters. This year the government allows the farmer to expand his area under cultivation by 20%. The farmer will extend the width and the length of the plowed area by equal amounts, x.

80 + x

120 m by 80 m

←x→

120 + x

14 Write an expression for the area of the new field in terms of x.

15 Equate this expression to the area of the new field; that is, the original area plus 20% of the original area. Combine and simplify terms to obtain a quadratic equation in the form $ax^2 + bx + c = 0$.

16 Solve the quadratic equation for x. How much wider and longer should the farmer plow the field to obtain a 20% increase in area?

An airplane drops feed bags to horses wintering in a mountain pasture. When dropped from a plane at 1000 feet above the horses, the approximate height h of the feed bags after t seconds is modeled by the quadratic equation $h = -16t^2 + 1000$.

17 How long will it take the feed bags to reach the ground? Give the time to the nearest tenth of a second.

18 Suppose the plane takes a second pass at 1500 feet. Write the equation to model the height h and the time t of the feed bags dropped from this altitude. How long will it take the feed bags to reach the ground?

BUSINESS & MARKETING

You have been instructed to get quotes on the cost for a billboard advertisement. One company determines the total cost from a formula based on an initial setup fee and the height (h) and width (w) of the sign. The company's specifications assume the width of the sign is 5 feet plus twice the height.

Total cost =
    Setup fee + (Cost per square foot) (Area of sign)

19 For the type of sign you need, the setup fee is $150 and the cost per square foot is $10. Substitute these values and the width specifications into the cost equation to obtain a quadratic equation in h, the height.

20 Your budget allows a total cost of no more than $4000 for the billboard. Set the Total cost equation equal to $4000 and solve the quadratic equation for the height (h) of the sign you can afford.

A manufacturing company uses the function $y = 81x - 81x^2$ to model the profit on one of its products. In the model, x represents the number of items in thousands.

21 What is the maximum profit the company can make on this product?

22 How many items must the company produce to make this profit?

23 What are the break-even points for production?

The recommended dosage of a certain type of medicine is determined by the patient's body weight. The formula to determine the dosage is

$$D = 0.1w^2 + 5w$$

where D is the dosage in milligrams (mg) and w is the patient's body weight in kilograms.

24 What is the maximum weight of a patient that you can treat with 1800 mg of this medicine?

A report on river contamination downstream from a waste treatment plant contains a graph, as shown below. The graph shows the amount of contaminant in parts per million (ppm), measured at various distances downstream from the plant. The distances are in units of thousands of feet. (Thus, 5 represents 5000 feet, 10 represents 10,000 feet, and so on.)

25 Which of the following quadratic equations best describes the trend shown by the data? In the equations, d represents the downstream distance in units of 1000 feet and c represents the concentration in parts per million.

a. $c = 2d^2 - 90d + 1000$
b. $c = -1000d^2 + 1000$
c. $c = d^2 + 20d - 1000$
d. $c = d^2 + 1000$

26 Use the equation that you selected to predict how far downstream the concentration drops to 10 ppm.

As a safety engineer for a chemical manufacturing company, you are designing a chemical waste holding area. It is located on a rectangular lot that is 200 yards long and 75 yards wide. The holding area must be 10,000 square yards. There must be a "safety zone" of uniform width around the perimeter of the holding area.

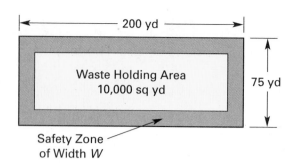

200 yd

Waste Holding Area
10,000 sq yd

75 yd

Safety Zone
of Width *W*

27 Let *W* represent the width of the safety zone. Write expressions for the length and the width of the interior holding area in terms of *W*.

28 Write an equation for the area of the interior holding region, using the specified area of 10,000 square yards and the product of the length and the width from Exercise 27.

29 Multiply the length and width terms and write a quadratic equation in the form $ax^2 + bx + c = 0$.

30 Use the quadratic formula to determine the width of the safety zone that makes your equation true.

FAMILY AND CONSUMER SCIENCE

The boiling point of water is affected by altitude. If you are boiling water at an altitude of *h* feet above sea level, it will not boil at exactly 212°F, but a few degrees lower. Since the water is not quite as hot, you may have to boil vegetables longer to completely cook them. The following equation approximates the number of degrees Fahrenheit below 212°F (the drop *D*) at which water will boil at various altitudes above sea level:

$$D^2 + 520\,D = h$$

31 Substitute 10,000 feet for *h* and rewrite the equation above in the standard form for a quadratic equation. For an altitude of 10,000 feet above sea level, about how much will the boiling point of water drop below 212°F?

32 What is the approximate boiling point at the altitude of your school?

You have just won a contract to select a wall hanging for the lobby of a new theater. The wall is 32 feet long, and the ceiling is 20 feet high. The theater owners want the wall hanging no closer than 2 feet from the ceiling, the floor, and each side of the wall. They also want the hanging to have the length and height dimensions of the ratio of the Golden Rectangle.

The Golden Rectangle has dimensions of length (L) and height (H) that satisfy the following formula:

$$\frac{L + H}{L} = \frac{L}{H}$$

33 Cross-multiply the terms in the proportion to obtain a quadratic equation. Substitute the values for either the maximum vertical distance available or the maximum horizontal distance available. Organize the resulting equation into the form $ax^2 + bx + c = 0$, where x represents the unknown dimension of the wall hanging.

34 Solve the quadratic equation for the unknown dimension. Determine whether there is enough room for a wall hanging that satisfies the ratio of the Golden Rectangle.

You are designing the layout for a square quilt: a patchwork of 3-inch squares with a solid piece as a backing. You need to determine how many squares will fit onto the backing material, which is 60 inches by 60 inches, or 3600 square inches.

The backing will be folded to make a $\frac{1}{2}$-inch border on each of the four edges of the quilt (see the enlarged side view in the illustration). This will use 1 inch of material for each edge. To obtain a quilt with rows of squares that total z inches long, you must allow for a backing that is $z + 2$ inches long (see the illustration). Each row of the quilt is made of 3-inch squares so that the length of a row is $3n$ inches, where n is the number of squares. Thus, $z = 3n$.

35 The area of the backing equals its length times its width. Write an equation for the area of the backing in terms of z, using the information and the drawing. (*Hint:* Remember that the final quilt will be square.)

36 The length z of a finished row of squares is a multiple of the 3-inch size of each square; that is, $z = 3n$. Substitute this expression for z into your equation from Exercise 35.

37 Write your equation in the standard form for a quadratic equation. Solve for n.

38 Based on the amount of backing material you have, what is the maximum number of 3-inch squares per row that you can use? (*Hint:* Remember that n must be a whole number.)

39 What is the length and the width of the backing material you will use?

INDUSTRIAL TECHNOLOGY

A BASIC computer program is used to help locate the roots of a quadratic equation. The program calculates y-values for a range of x-values that is input by the user. The program prints the values as ordered pairs. Use the computer output shown to answer each of the following.

40 What range of x-values was input by the user?

41 How would you use this output to identify the root(s) for the equation under study?

42 Estimate the root(s) of the equation.

43 How could you modify the program to increase the accuracy of your estimate?

44 Since the equation is quadratic, it has a minimum or a maximum y-value. Which one does the output show: a maximum or a minimum? Estimate the maximum (or minimum).

X	Y		X	Y
(−2	, 1.331)		(0	, −1.969)
(−1.9	, .976)		(.1	, −1.924)
(−1.8	, .641)		(.2	, −1.859)
(−1.7	, .326)		(.3	, −1.774)
(−1.6	, .031)		(.4	, −1.669)
(−1.5	, −.244)		(.5	, −1.544)
(−1.4	, −.499)		(.6	, −1.399)
(−1.3	, −.734)		(.7	, −1.234)
(−1.2	, −.949)		(.8	, −1.049)
(−1.1	, −1.144)		(.9	, −.844)
(−1.	, −1.319)		(1	, −.619)
(−.9	, −1.474)		(1.1	, −.374)
(−.8	, −1.609)		(1.2	, −.109)
(−.7	, −1.724)		(1.3	, .176)
(−.6	, −1.819)		(1.4	, .481)
(−.5	, −1.894)		(1.5	, .806)
(−.4	, −1.949)		(1.6	, 1.151)
(−.3	, −1.984)		(1.7	, 1.516)
(−.2	, −1.999)		(1.8	, 1.901)
(−.1	, −1.994)		(1.9	, 2.306)
			(2	, 2.731)

In an electric circuit, the power dissipated in a load is given by the following equation.

$$P = VI - I^2R$$

P is the power used by the load in watts.
V is the voltage supplied in volts.
R is the resistance of the circuit external to the load in ohms.
I is the circuit current in amperes.

45 In a certain circuit, the voltage supplied is 110 volts, and the resistance of the circuit external to the load is 30 ohms. What is the current if the load uses 60 watts of power?

You need to construct a rain gutter from a rectangular piece of sheet metal. The sheet metal is 20 inches wide. To form the gutter, you will fold up two sides of equal width to form right angles, as shown. For proper drainage, the gutter needs to have a cross-sectional area of 38 in².

46 Write an expression for the width of the gutter, in terms of the height x of the gutter's sides.

47 Write an equation for the cross-sectional area of the gutter, in terms of the width from Exercise 46 and the height x. Your equation should satisfy the requirements for a cross-sectional area of 38 square inches. Rewrite the equation in the standard form for a quadratic equation.

48 Use the quadratic formula or the method of completing the square to determine the values of the height of the gutter.

49 Sketch a scale drawing that shows the two solutions obtained in Exercise 48. Which solution would you choose for the gutter height?

You are designing an archway that has a radius of 25 feet and spans a distance of 40 feet. The formula below gives the relationship between the center height (*h*) of the arch, the radius of curvature (*R*) of the arch, and the distance (*s*) spanned by the arch.

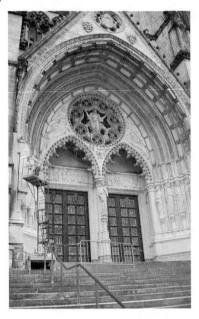

$$s^2 = 8Rh - 4h^2$$

50 Substitute the given values to obtain a quadratic equation for the center height *h*.

51 Solve the quadratic equation. Which value of *h* would you use for the center height of your arch? Why did you choose this value?

You are selecting a piece of angle iron for a construction project. The weight of the angle iron will exceed specifications if it has a cross-sectional area greater than 60 square centimeters. You need a vertical height of 16 cm and a horizontal width of 10 cm. Find the thickness of the stock *x* that satisfies these conditions.

52 The dotted line divides the angle iron's cross section into two rectangles, each with a width of *x*. Write an expression for the length of each rectangle. Multiply length and width to write equations for the area of each rectangle.

53 Add the two areas and write an equation for the total cross-sectional area of the angle iron. Set this equation equal to the maximum area of 60 square centimeters.

54 Rewrite the equation in standard form and solve for x. Which value of x would you choose for the thickness of the angle iron?

A cylindrical tank is placed along a wall, as shown in the side-view drawing below. The tank has a diameter ($2R$) of 84 inches. You need to run a pipe along the wall between the wall and the tank. You need to determine the radius r of the largest pipe that can fit in the space left by the tank.

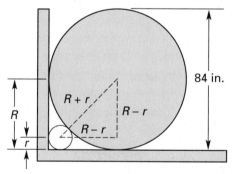

55 A right triangle is shown by dotted lines. The hypotenuse of the triangle is formed by the line that joins the centers of the two circular cross sections and has a dimension of $R + r$. Each side of the triangle is equal to the radius of the tank (R) less the radius of the pipe (r), or $R - r$. Use the Pythagorean Theorem ($c^2 = a^2 + b^2$) with these dimensions to obtain a quadratic equation for r and R. Substitute for R and write the equation in standard form.

56 Solve the quadratic equation for r to find the largest pipe that will fit between the tank and the wall.

You want to increase the weight of an iron counterweight by 20% by bolting two iron plates (of the same type of iron) along the top and the side. The weight of the iron stock is proportional to its cross-sectional area. Find the thickness x of iron plate stock needed to increase the overall cross-sectional area by 20%.

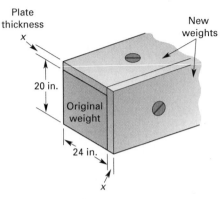

57 The original counterweight is 20 inches by 24 inches and its cross-sectional area is 480 square inches. If you increase this area by 20%, what is the new cross-sectional area?

58 The new cross-sectional area will be the product of the new height and the new width. Refer to the drawing. Write an expression for the new area in terms of the plate thickness x. Equate this expression to the larger area from Exercise 57.

59 Rewrite the equation for the new area in standard form and solve for the plate thickness x.

Most three-way light bulbs have only two filaments. For example, one filament may be 50 watts, and the second filament may be 100 watts. The bulb produces three levels of brightness by using the first filament (50 watts), the second filament (100 watts), or both filaments together (150 watts). However, your eye will not see equal ratios of brightness for these three power levels. Switching from the first to the second level yields a brightness ratio of 100:50, or 2:1. However, going from the second to the third level yields a ratio of 150:100, or 3:2.

Suppose you are asked to design a three-way bulb that has a 50-watt element as its first filament and yields equal ratios for 2nd level:1st level and 3rd level:2nd level.

Let x be the wattage of the second filament. The wattage of the third level will be $x + 50$. You want the following relationship:

$$\frac{2\text{nd level}}{1\text{st level}} = \frac{3\text{rd level}}{2\text{nd level}}$$

or

$$\frac{x}{50} = \frac{x + 50}{x}$$

60 Find a quadratic equation for x by cross multiplying the terms in this proportion.

61 Use the quadratic formula to determine the wattage x for the second filament.

62 Check the results from Exercise 61. What is the ratio of the second-to-first-filament wattages? What is the ratio of the third-to-second-filament wattages? Are the ratios equal?

63 Suppose the first filament is 40 watts. Find the wattage of the second filament.

64 Now let the first filament be any power level W. Show that as long as the ratios of 2nd level:1st level and 3rd level:2nd level are equal, the ratios are equal to 1.62.

Skills

Graph each function. Give the vertex of the graph.

1. $y = x^2 + 3$

2. $y = (x - 1)^2 + 3$

Let $f(x) = x^2 - 3x + 2$.

3. Find the zeros of the function by factoring.

4. Write the function in parabola form.

5. Give the coordinates of the vertex.

6. Write the equation for the axis of symmetry.

Solve each equation using the quadratic formula.

7. $6x^2 + 16x + 9 = 0$

8. $-20x^2 + 9x - 1 = 0$

9. Use the discriminant to find the number of real roots of $y = x^2 + 5x + 8$.

Applications

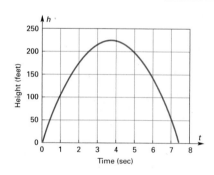

The parabola describes a ball's height versus time of flight.

10. Estimate the maximum height of the ball.

11. Estimate the time the ball stays in the air.

12. The sum of the first n whole numbers, $1 + 2 + 3 + \cdots + n$, is determined by the formula

$$\text{Sum} = \frac{n^2 + n}{2}$$

Suppose the sum is 2628. Find the value of n.

Your garden measures 16 feet by 48 feet. You want to reduce its area of 768 square feet to 585 square feet. However, you need to reduce the length and the width by the same number of feet.

Original boundary of garden

13. Write a quadratic equation to model the situation.

14. By how many feet will you reduce the length and the width of the original garden?

15. The formula that relates the center height (h) of an arch, the radius of curvature (R), and the distance spanned (s) is

$$s^2 = 8Rh - 4h^2$$

You are designing an archway to span a 36-foot distance. The arch must have a radius of curvature of 24 feet. Find the center height to the nearest tenth of a foot.

Math Lab

16. What are the coordinates of the focus of the parabola shown at the left?

17. Write an equation for the cross-sectional area of a gutter constructed from a piece of sheet metal 12 inches wide with sides that are equal to each other. (*Hint:* Sketch the gutter cross section.)

18. You are using a computer program to solve a quadratic in the form $y = ax^2 + bx + c$. You input upper and lower values for x, and the program calculates values for y. When you run the program, it prints the following:

x	y
5.0	−2.33
5.1	−1.55
5.2	−0.27
5.3	0.42
5.4	1.84
5.5	2.96

Estimate one of the roots of the quadratic.

WHY SHOULD I LEARN THIS?

Describing geometric figures with algebra is an essential workplace skill. Architects use algebra to design the shape of a sloped roof or determine the spacing between rafters. Carpenters transform architectural designs into buildings by using several tools including algebra. If designing or building ever become part of your career goals, you will use the concepts in this chapter.

RIGHT TRIANGLE RELATIONSHIPS

OBJECTIVES

1. Name the parts of a right triangle.
2. Use the Pythagorean Theorem to find a side of a right triangle.
3. Use the characteristics of 3-4-5, 45°-45°, and 30°-60° right triangles to solve problems.
4. Use the ratios for the sine, cosine, and tangent of an angle to solve problems that involve right triangles.
5. Add, subtract, and simplify expressions containing radicals.

Look around you. How many triangles can you see? Triangles are everywhere. Architects use triangles to brace corners, add strength to buildings, and support signs. Triangles are used to construct homes, offices, dams, and towers. They are also used in navigation to help airplanes and ships find their exact positions.

People have known for a long time that triangles are important and useful. Long before books were printed or motors invented, people knew about the mathematics of triangles. The Greeks, more than 2500 years ago, worked out most of the mathematics that use properties of triangles to solve problems. You will use these properties to solve many workplace problems.

In this chapter, you will study several properties of right triangles. As you watch the video, notice how people use right triangles on the job.

LESSON 12.1 SIMILAR TRIANGLES

Before you begin working with triangles, review this list of general properties of triangles. They will be used throughout this chapter.

General Properties of Triangles

a. Triangles have 3 sides and 3 angles.

b. $m\angle A + m\angle B + m\angle C = 180°$

c. Perimeter $= l + m + n$

d. Area $= \frac{1}{2}bh$

- All triangles have three sides and three angles. The three sides lie in the same plane.

- The sum of the measures of the three angles of any triangle is equal to 180°.

- The perimeter of a triangle is equal to the sum of the lengths of the three sides.

- The area of a triangle is equal to the product of one-half the base times the height.

Triangles are labeled in a special way. Examine the information below to refresh your memory.

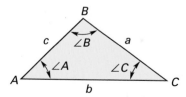

- The three vertices of a triangle are often labeled with capital letters, such as A, B, and C.
- The triangle is referred to as $\triangle ABC$.
- The sides opposite the angles are often labeled with lowercase letters, such as a, b, and c. Note that $a =$ length of BC, $b =$ length of AC, and $c =$ length of AB.
- The three angles of the triangle are often labeled with the appropriate vertex letter, using the symbol $\angle$ for angle, such as $\angle A$, $\angle B$, and $\angle C$.

Right Triangles

A right triangle has one **right angle** with a measure of 90°. The sides that form the right angle are perpendicular to each other.

A right triangle has one angle equal to 90°.

You can divide any triangle into two right triangles. For example, Triangle *ABC*, written Δ*ABC*, is *not* a right triangle. But by drawing a perpendicular line segment from vertex *B* to point *M* on side *AC*, you can form two right triangles, Δ*ABM* and Δ*CBM*.

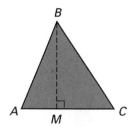

This illustration labels the parts of a right triangle.

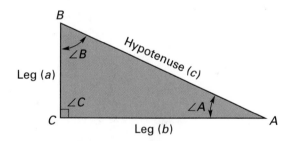

The following list summarizes the important facts about a right triangle.

1. A right triangle has one right angle. Angle *C* is the right angle in triangle *ACB*. The measure of angle *C* is 90 degrees. This is written

$$m\angle C = 90°$$

2. The side opposite the right angle is the longest side and is called the **hypotenuse.** The hypotenuse is labeled as *c*.

3. The two sides that meet at right angles are called the **legs** of the right triangle. The length of the leg opposite *A* is labeled *a*, and the length of the leg opposite *B* is labeled *b*.

Similar Right Triangles

Similar Right Triangles

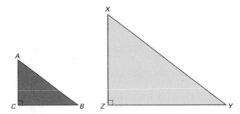

1 Use a protractor to measure each angle to the nearest whole degree in the two right triangles. Complete the following information.

$m\angle C = 90°$	$m\angle Z = 90°$
$m\angle A = ?$	$m\angle X = ?$
$m\angle B = ?$	$m\angle Y = ?$

2 What is the relationship between $m\angle A$ and $m\angle X$? $m\angle B$ and $m\angle Y$?

3 Use a ruler to measure the length of each side of the two triangles. Copy and complete this information.

length of $AB = ?$	length of $XY = ?$
length of $AC = ?$	length of $XZ = ?$
length of $BC = ?$	length of $YZ = ?$

4 Write each ratio in decimal form.

$$\frac{AB}{XY} \qquad \frac{AC}{XZ} \qquad \frac{BC}{YZ}$$

5 Summarize your findings.

You found in Step 1 that the three angles of $\triangle ABC$ are equal respectively to the three angles of $\triangle XYZ$. Therefore, $\triangle ABC$ is *similar* to $\triangle XYZ$. The three pairs of equal angles are called **corresponding angles.** The sides opposite the corresponding angles are called **corresponding sides.** You found in Step 4 that the ratios of corresponding sides of similar triangles are equal.

Similar Triangles

Two triangles are similar if they have corresponding angles with the same measure.

If two triangles are similar, then the lengths of their corresponding sides are proportional.

Similar triangles have the same shape but not necessarily the same size. Model builders use the properties of similar triangles to build scale models.

EXAMPLE 1 Finding a Missing Side

$\triangle MNO$ is similar to $\triangle QRS$.

Find the length of side QS.

SOLUTION

Since the two triangles are similar, the lengths of the corresponding sides are proportional. Write a proportion and solve.

$$\frac{MN}{QR} = \frac{MO}{QS}$$ Similar Triangle Proportion

$$\frac{4}{6} = \frac{6}{a}$$ Substitution

$$4a = 36$$ Proportion Property

$$a = 9$$ Division Property of Equality

The length of QS is 9 feet.

Ongoing Assessment

For $\triangle MNO$ and $\triangle QRS$, change the lengths to the following: Let the length of RS equal 15 inches, the length of MO equal 6 inches, and the length of NO equal 9 inches. Find the length of QS.

EXAMPLE 2 Indirect Measurement

How can a surveyor find the distance across the lake from point A to point D?

SOLUTION

The surveyor can draw two right triangles, $\triangle ABC$ and $\triangle AED$. The triangles share a common angle, $\angle A$. Both $\angle C$ and $\angle D$ are $90°$. Thus, $m\angle B = m\angle E$. (Can you show why?) Since $\triangle ABC$ and $\triangle AED$ have three equal angles, they are similar triangles.

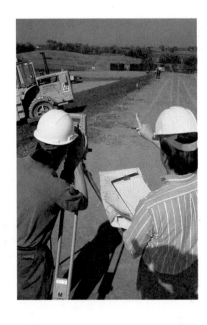

The surveyor can measure the distances CB, DE, and CD on land. Then the surveyor can use the similar triangle property to write the following ratios:

$$\frac{AD}{AC} = \frac{DE}{CB} \qquad \text{Similar Triangle Property}$$

If the distance AD across the lake is represented by x, the ratios become

$$\frac{x}{x + 40} = \frac{80}{120} \qquad \text{Substitution}$$

$$120x = 80x + 3200 \qquad \text{Proportion Property; Distributive Property}$$

$$40x = 3200 \qquad \text{Subtraction Property}$$

$$x = 80 \qquad \text{Division Property}$$

The distance across the lake is 80 meters.

LESSON ASSESSMENT

Use the diagram for Exercises 1 and 2.

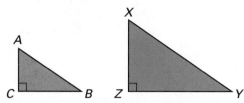

1 Suppose that angle *A* has the same measure as angle *X*. Explain why the two triangles are similar.

2 Suppose the length of *AC* is 6 inches, the length of *CB* is 9 inches, and the length of *XZ* is 10 inches. Explain how to find the length of *ZY*.

Practice and Problem Solving

Each pair of triangles is similar. Find the length of each missing side.

3.

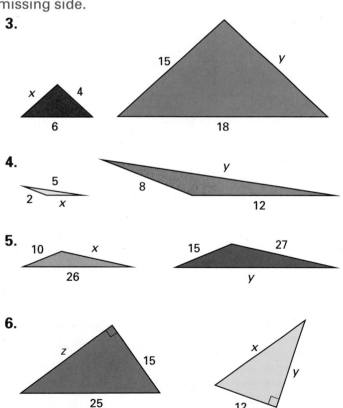

4.

5.

6.

Solve each problem. Draw a diagram when needed.

7. The length of *DE* is 40 yards, the length of *CB* is 180 yards, and the length of *BE* is 50 yards. How far is it across the lake from point *C* to point *A*?

$\triangle BED \sim \triangle BCA$

8. Suppose the length of *DE* is 4 meters, the length of *AD* is 8 meters, and the length of *BC* is 6 meters. What is the length of *AC*?

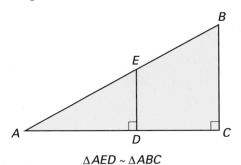

$\triangle AED \sim \triangle ABC$

9. The lengths of the three sides of a triangle are 10 inches, 15 inches, and 20 inches. The longest side of a similar triangle is 25 inches. Find the lengths of the other two sides.

10. A tree casts a 25-foot shadow at the same time that an upright yardstick casts a 10-foot shadow. Use similar triangles to find the height of the tree.

11. The top of a 30-foot ladder touches the side of a building at a point 25 feet above the ground. A 12-foot ladder is placed against the building at the same angle. How far from the ground is the top of the 12-foot ladder?

For Exercises 12–15, use the following ordered pairs:

$A(-3,1)$ $B(1,1)$ $C(1,-2)$ $D(-3,-2)$

12. Graph the four points on the same coordinate plane.

13. Find the slope of a line containing AD and a line containing BC.

14. Write the equation of a line containing AB.

15. Write the equation of a line containing CD.

Graph the solution to each inequality.

16. $4x + 3 > -15$ and $2x < 6$

17. $2x - 1 \le 5$ or $-x > -3$

18. $|2x + 1| \le 9$

Write and solve an inequality for each statement.

19. The length of a missile is ten times the length of its nose cone. The difference in length between the missile and its nose cone is at least 16 feet. What is the length of the missile?

20. The sum of two acute angles in a triangle is less than or equal to 90°. One angle is twice the measure of the other. What is the greatest possible measure of the smaller angle?

LESSON 12.2 THE PYTHAGOREAN THEOREM

Recall the special relationship between the sides of a right triangle that you learned in Chapter 6.

In a right triangle, the square of the length of the hypotenuse equals the sum of the squares of the lengths of the two legs. Although many civilizations knew and used this fact, it is now known as the **Pythagorean Theorem.** The Pythagorean Theorem states that for any right triangle:

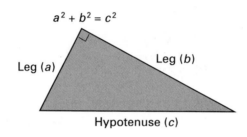

$$a^2 + b^2 = c^2$$

Leg (a)   Leg (b)

Hypotenuse (c)

ACTIVITY 1   Pythagorean Triples

A set of three positive integers that satisfy $a^2 + b^2 = c^2$ is called a **Pythagorean triple.** For example, the set 3, 4, and 5 is a Pythagorean triple because

$$3^2 + 4^2 = 5^2$$

That is, $9 + 16 = 25$.

Complete each set of integers to form a Pythagorean triple.

		a	b	c
1	Triangle One	7	24	?
2	Triangle Two	8	15	?
3	Triangle Three	?	63	65
4	Triangle Four	28	?	53

The Sides of a Right Triangle

In Chapter 6, the Pythagorean Theorem provided a way to find the length of a leg or the hypotenuse of a right triangle. The Pythagorean Theorem is also a tool for determining if a triangle is a right triangle. If you measure the three sides of a triangle and find that the sum of the squares of two sides is equal to the square of the longest side, then the triangle is a right triangle.

EXAMPLE 1 Checking for a Right Triangle

A 12-foot antenna pole is held in place by two guy wires of unequal length. One end of each guy wire is connected to the top of the pole. The shorter guy wire is 15 feet long. It is fastened to the ground 9 feet from the base of the pole. Does the pole form a right angle with the ground?

SOLUTION

First, sketch the situation to see the position of the pole and the wires.

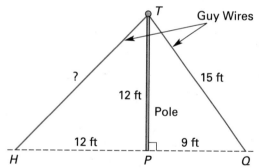

Label the triangle and its sides. Substitute the values in the formula $a^2 + b^2 = c^2$. Does 15^2 equal $12^2 + 9^2$?

Since $225 = 144 + 81$, $\triangle TPQ$ is a right triangle, and the pole forms a right angle with the ground.

Ongoing Assessment

Look at Example 1 again. Suppose you fasten the second guy wire to the ground at point H, 12 feet from the base of the pole. Find the length of the guy wire to the nearest foot.

An electrical circuit wired in series contains an alternating voltage source, a resistor, and a coil. The voltage across the resistor (V_R) and the voltage across the coil (V_L) are related to the voltage source (V_S) by the following formula:

$$(V_R)^2 + (V_L)^2 = (V_S)^2$$

Notice that this relationship is the same as the relationship between the sides of a right triangle. Because of this, these voltages are often represented in a voltage triangle.

a. Circuit   **b.** Voltage triangle

EXAMPLE 2 An Electrical Circuit in Series

In an electrical circuit such as the one above, the voltage source is 185 volts. The voltage across the resistor is 76 volts. Find the voltage across the coil to the nearest tenth.

SOLUTION

First, sketch a voltage triangle to help you see the situation.

Next, label the hypotenuse and legs of the triangle.

Substitute the values in the Pythagorean Theorem and solve for the unknown voltage.

$$(V_R)^2 + (V_L)^2 = (V_S)^2$$

$$(76)^2 + (V_L)^2 = (185)^2$$

$$(V_L)^2 = (185)^2 - (76)^2$$

$$(V_L)^2 = 34{,}225 - 5776$$

$$(V_L)^2 = 28{,}449$$

$$(V_L) = \sqrt{28{,}449}$$

$$(V_L) \approx 168.7$$

The voltage across the coil is about 168.7 volts.

Firefighters in training practice a rescue operation from a building. Their ladder is 20 feet long, and they place the foot of the ladder 6 feet from the base of the building. How high up the wall will the top of the ladder reach?

The Distance Formula

ACTIVITY 2   Distance on the Coordinate Plane

1   In a coordinate plane, draw a line segment from A (3,2) to B (6,4).

2   Graph point C (6, 2). Draw $\triangle ACB$. Explain why $\triangle ACB$ is a right triangle.

3   Find the length of AC and BC.

4   Use the Pythagorean Theorem to find AB to the nearest tenth.

5   Explain how to use the Pythagorean Theorem to find the distance between M (2, 3) and N (5,7).

This Activity demonstrates that you can use the Pythagorean Theorem to find the distance between any two points on the coordinate plane.

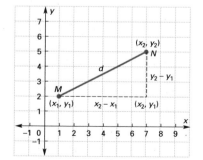

The Distance Formula

Let $M(x_1, y_1)$ and $N(x_2, y_2)$ be any two points on the Cartesian coordinate plane. The distance (d) between the two points M and N is

$$d = \sqrt{(x_2 - x_1)^2 + (y_2 - y_1)^2}$$

Critical Thinking Explain why you can choose either point M or N for (x_1, y_1) in the distance formula.

EXAMPLE 3 Applying the Distance Formula

Two ships are several miles apart. A ship's navigator plots their locations using coordinates. One ship is located at P (30,50). The other ship is located at Q (105,150). What is the distance between the ships?

SOLUTION

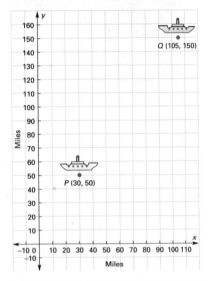

Draw a diagram. Use the distance formula.

$$d = \sqrt{(105 - 30)^2 + (150 - 50)^2}$$

$$d = \sqrt{(75)^2 + (100)^2}$$

$$= \sqrt{15,625}$$

$$= 125$$

The ships are 125 miles apart.

LESSON ASSESSMENT

Think and Discuss

1 The lengths of the three sides of a triangle are represented by a, b, and c. Under what condition is the triangle a right triangle?

2 Explain how to find the distance between two points on a coordinate plane.

Practice and Problem Solving

Each set of integers represents the lengths of the sides of a triangle. Use the Pythagorean Theorem to determine whether the triangle is a right triangle. Answer *yes* or *no*.

3. 5, 12, 15   **4.** 24, 7, 25   **5.** 17, 15, 8

6. 12, 37, 35   **7.** 81, 82, 18   **8.** 30, 16, 34

Use a diagram to help solve each problem. When necessary, round the answer to the nearest tenth.

9. A rectangular swimming pool is 9 yards wide and 20 yards long. A person swims diagonally across the pool from one corner to the other. How far does the person swim?

10. A lawn is in the shape of a rectangle. A water pipe 45 feet long is buried diagonally across the lawn. The length of the lawn is 25 feet. How wide is the lawn?

11. Two telephone poles are 50 feet apart. The height of one pole is 50 feet. The height of the other pole is 55 feet. What is the minimum length of wire that can be strung between the tops of the two poles?

12. A carpenter is building a rectangular gate 4 feet by 6 feet. She needs to cut a metal rod to brace the gate diagonally. How long should she cut the brace?

Find the distance between each pair of points.

13. $A(4, -1)$ and $B(7, 3)$ **14.** $C(2, -3)$ and $D(-1, 3)$

15. $M(-5, 9)$ and $N(3, -6)$ **16.** $X(0, 1)$ and $Y(-1, 2)$

Three ships are located at the following points: $A(4, -1)$; $B(4, 11)$; and $C(8, -1)$.

17. Prove the ships are at the corners of a right triangle.

18. Find the perimeter of the triangle.

Mixed Review

A flooring contractor is tiling a kitchen floor that is 9 feet by 12 feet. The tile costs $2 per square foot. The tile cement costs $4.50 for a gallon can and covers 10 square yards.

19. What is the cost of the kitchen floor tile?

20. How many square yards of cement are needed?

21. How many gallon cans of cement must be purchased? What is the cost of the cement?

22. Write a formula that can be used to find the total cost of the tile and cement for a floor x feet by y feet.

LESSON 12.3 IRRATIONAL NUMBERS AND RADICALS

Solving problems with right triangles requires skill in working with **radicals.** Often, when you evaluate radicals, the result is an **irrational number**. For example, $\sqrt{7}$ is an irrational number. You cannot write $\sqrt{7}$ in the form $\frac{a}{b}$ where $b \neq 0$ and a and b are integers.

Recall that the symbol $\sqrt{}$ is a radical sign. The expression under the radical sign is called the **radicand**. The expression $\sqrt{7}$ is read "radical seven."

Simplifying Radicals

All Pythagorean triples are integers. However, not all numbers that make the Pythagorean Theorem true are integers. For example, the right triangle in this illustration has two legs that measure 2 centimeters on a side. What is the length of the hypotenuse?

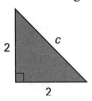

If you use the Pythagorean Theorem, you will find the following:

$$c^2 = (2)^2 + (2)^2$$
$$c^2 = 8$$
$$c = \pm \sqrt{8}$$

Since length is always a positive number, the hypotenuse is $\sqrt{8}$ centimeters.

The number $\sqrt{8}$ is an irrational number. You cannot write $\sqrt{8}$ as a decimal or a fraction. However, you can use your calculator to approximate $\sqrt{8}$.

Rounded to the nearest thousandth, $\sqrt{8}$ is 2.828.

You have seen that a Pythagorean triple cannot represent a right triangle with legs 2 centimeters in length. In fact, the lengths of the sides of most triangles cannot be described by a combination of three integers. At least one of the sides is usually an irrational number. Thus, most of the work you do with right triangles will involve irrational numbers written as radicals.

Use your calculator to approximate each expression to the nearest thousandth.

	A	B	C	D
1	$\sqrt{64}$	$\sqrt{16 \cdot 4}$	$\sqrt{16} \cdot \sqrt{4}$	$4 \cdot 2$
2	$\sqrt{32}$	$\sqrt{16 \cdot 2}$	$\sqrt{16} \cdot \sqrt{2}$	$4\sqrt{2}$
3	$\sqrt{125}$	$\sqrt{25 \cdot 5}$	$\sqrt{25} \cdot \sqrt{5}$	$5\sqrt{5}$
4	$\sqrt{27}$	$\sqrt{9 \cdot 3}$	$\sqrt{9} \cdot \sqrt{3}$	$3\sqrt{3}$

5 Examine the relationship between column B and column D. What is the relationship between the square root of a product of two numbers and the product of their square roots?

This activity demonstrates two important properties of radicals that simplify your work with irrational numbers.

> **Calculating with Radicals**
> If a and b are any two numbers,
> $$\sqrt{a} \cdot \sqrt{b} = \sqrt{ab}$$
> If a and b are any two numbers with $b \neq 0$,
> $$\sqrt{\frac{a}{b}} = \frac{\sqrt{a}}{\sqrt{b}}$$

A radical is in **simplest form** when the radicand does not contain a factor (other than one) whose square root is a whole number. For example, $\sqrt{32}$ is not in simplest form because one factor of 32 is 16 and the square root of 16 is 4. However, $\sqrt{32}$ can be written as $4\sqrt{2}$. The expression $4\sqrt{2}$ is in simplest form because $\sqrt{2}$ does not contain a factor whose square root is a whole number.

EXAMPLE 1 Multiplying Radicals

Write each of the following in simplest form.

a. $\sqrt{6} \cdot \sqrt{8}$ **b.** $\sqrt{8} \div \sqrt{2}$

SOLUTION

a. $\sqrt{6} \cdot \sqrt{8} = \sqrt{48}$ 　　　　**b.** $\sqrt{8} \div \sqrt{2} = \sqrt{4}$
$\qquad\qquad\quad = \sqrt{16} \cdot \sqrt{3}$ 　　　　　　　　$= 2$
$\qquad\qquad\quad = 4\sqrt{3}$

Write each of the following in simplest form.

a. $\sqrt{12} \cdot \sqrt{6}$ **b.** $\sqrt{50} \div \sqrt{2}$

When dividing radicals, it is customary to write the denominator without a radical. For example, to eliminate the radical in the denominator of $\frac{3}{\sqrt{6}}$, multiply $\frac{3}{\sqrt{6}}$ by 1, in the form $\frac{\sqrt{6}}{\sqrt{6}}$.

The result is

$$\frac{3}{\sqrt{6}} \cdot \frac{\sqrt{6}}{\sqrt{6}} = \frac{3\sqrt{6}}{\sqrt{36}}$$

$$= \frac{3\sqrt{6}}{6}$$

$$= \frac{\sqrt{6}}{2}$$

This process is called **rationalizing** the denominator.

CULTURAL CONNECTION

The mathematicians living in Alexandria, Egypt, about 75 BCE discovered a way to find the area (A) of any triangle given only the length of each of the sides. Represent the lengths of the sides of a triangle by a, b, and c. Let s equal $\frac{1}{2}$ the perimeter of the triangle; that is, $s = \frac{1}{2}(a + b + c)$.

The following area formula is named for Heron of Alexandria.

$$A = \sqrt{s(s - a)(s - b)(s - c)}$$

To find the area of a triangle with sides of 6 inches, 9 inches, and 9 inches, first find s.

$$s = \frac{1}{2}(6 + 9 + 9) = 12$$

Substitute the numerical values into Heron's Formula.

$$A = \sqrt{12(12 - 6)(12 - 9)(12 - 9)}$$

$$= \sqrt{(12)(6)(3)(3)}$$

$$= \sqrt{36 \cdot 9 \cdot 2}$$

$$= \sqrt{36} \cdot \sqrt{9} \cdot \sqrt{2}$$

$$= 6 \cdot 3 \cdot \sqrt{2}$$

$$= 18\sqrt{2}$$

The area of the triangle is $18\sqrt{2}$, or about 25.46 square inches.

1. Use Heron's Formula to find the area of a triangle with sides that measure 8 meters, 12 meters, and 10 meters.

2. Find the area of a triangle with sides that measure 8 inches, 12 inches, and 4 inches. Describe the shape of this triangle. Draw the triangle to check your thinking.

LESSON ASSESSMENT

Think and Discuss

1 How do you know when a radical is in its simplest form?

2 How do you rationalize a fraction that contains a radical in the denominator?

Practice and Problem Solving

Write in simplest form.

3. $\sqrt{12}$　　　**4.** $\sqrt{75}$　　　**5.** $\sqrt{200}$

6. $\sqrt{72}$　　　**7.** $\dfrac{4}{\sqrt{24}}$　　　**8.** $\sqrt{6} \cdot \sqrt{15}$

9. $\sqrt{90} \div \sqrt{5}$　　　　**10.** $\sqrt{6}(\sqrt{6} + \sqrt{18})$

The formula for the area (A) of a circle with radius r is $A = \pi r^2$.

11. Solve the formula for r.

12. Write a radical expression for r in simplest form for a circle with an area of 72 square feet.

You can approximate the velocity (in feet per second) of a tidal wave in water of depth d (in feet) with the formula $V = \sqrt{15d}$.

13. Write a radical expression in simplest form for the velocity of a tidal wave in 400 feet of water.

14. Find the approximate velocity of the tidal wave.

Graph the following functions on the same coordinate plane.

15. $y = \sqrt{x}$　　　**16.** $y = \sqrt{2x}$　　　**17.** $y = \sqrt{\left(\frac{1}{2}\right)x}$

18. Draw the graph of $y = \sqrt{2x} + 2$. What affect does adding a positive number to $\sqrt{2x}$ have on the graph of $y = \sqrt{2x}$?

Write an equation in one unknown to model each situation. Solve the equation and check the solution.

19. The width of a rectangular swimming pool is 25 feet less than its length. The distance around the pool is 250 feet. Find the dimensions of the swimming pool.

20. Two guy wires are attached to a TV transmitter. The wires form a triangle with the ground as its base. The base angles are equal, and each base angle is four times the third angle. What is the measure of each base angle?

Write a system of equations to model each situation. Solve the system and check the solution.

21. Althea is installing an underground sprinkler system. She needs to cut a 24-foot water pipe into two sections. The longer pipe must be 6 feet longer than twice the length of the shorter one. How long are the two sections?

22. Keiko ordered $427.50 worth of books in July. She ordered $382.50 worth in August. Keiko ordered the same number of fiction books and the same number of nonfiction books each month, but the number of fiction books was not the same as the number of nonfiction books. How many of each did Keiko order?

July and August Book Prices for		
	Fiction	Nonfiction
July	$8.50	$6.50
Aug	$7.50	$6.00

LESSON 12.4 SPECIAL RIGHT TRIANGLES

The 3-4-5 Ratio

Sometimes you can find the length of a side of a right triangle without directly using the Pythagorean Theorem. The ratios of the sides of three special right triangles can simplify your calculations.

ACTIVITY 1   The 3-4-5 Right Triangle

Four triangles with sides a, b, and c have the following lengths:

	a	b	c
Triangle 1	3	4	5
Triangle 2	6	8	10
Triangle 3	15	20	25
Triangle 4	18	24	30

1 Use the Pythagorean Theorem to show that each triangle is a right triangle.

2 Use the definition of similar triangles to show that each pair of triangles is similar.

3 For each triangle, find the greatest common factor m of a, b, and c.

4 Write the ratio of the lengths of each triangle as

$$a:b:c = x \cdot m:y \cdot m:z \cdot m$$

For example, since the greatest common factor for Triangle 1 is 1. The ratio is written:

$$3:4:5 = 3 \cdot 1:4 \cdot 1:5 \cdot 1$$

5 What do you notice about x, y, and z for each of these similar right triangles?

The triangles in Activity 1 are called **3-4-5 right triangles**.

3-4-5 Right Triangle

Every 3-4-5 right triangle has sides whose ratios are 3 to 4 to 5. Any triangle with sides of these ratios is a right triangle.

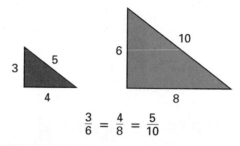

$$\frac{3}{6} = \frac{4}{8} = \frac{5}{10}$$

EXAMPLE 1 Using a 3-4-5 Right Triangle

A ship gathering specimens for marine biologists starts from an island and sails 100 miles due north. Then the ship turns and sails 75 miles due east. At that point, what is the shortest distance from the ship to the island?

SOLUTION

Draw a sketch to identify the right angle and the hypotenuse.

Notice that 100 and 75 have a greatest common factor of 25. Thus, you can write the ratio of the two legs as

$$\frac{75}{100} = \frac{3(25)}{4(25)}$$

Since the two legs of the right triangle have the ratio of 3 to 4, the three sides of the right triangle have the ratio of 3 to 4 to 5. To find the length of the hypotenuse, use the ratio and the greatest common factor, 25.

$$3{:}4{:}5 = 3 \cdot 25{:}4 \cdot 25{:}5 \cdot 25$$

The shortest distance back to the island is 125 miles.

To determine if a right triangle is a 3-4-5 right triangle, find the greatest common factor of the lengths of the sides. Then see if you can write the known sides as a ratio of

$$3 \cdot m : 4 \cdot m : 5 \cdot m$$

where m is the greatest common factor.

The side written as $5 \cdot m$ is always the longest side or hypotenuse.

The 45° - 45° Right Triangle

A triangle with two equal sides is called an **isosceles triangle.** In an isosceles triangle, the angles opposite the equal sides are also equal. If an isosceles triangle contains a right angle, it is called a **45°- 45° right triangle.**

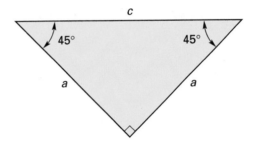

What happens when you apply the Pythagorean Theorem to the 45°- 45° right triangle? Since the two legs are equal in length,

$$a^2 + a^2 = c^2$$

$$2a^2 = c^2 \qquad \text{Combine like terms}$$

Solve the second equation for c.

$$c^2 = 2a^2$$

$$c = \sqrt{2}a \quad \text{or} \quad a\sqrt{2}$$

These steps show one unique characteristic of 45°- 45° right triangles.

> ### 45°- 45° Right Triangle
> For every 45°- 45° right triangle, the length of the hypotenuse is $\sqrt{2}$ times the length of a leg.

This drawing shows the main characteristics of a 45°- 45° right triangle.

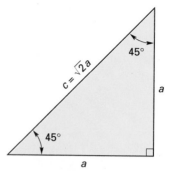

You can use a 45°- 45° right triangle to find the diagonal of a square.

Since the sides of a square are equal, the diagonal is the hypotenuse of a right triangle with two legs of equal length. Thus, the two right triangles formed by the diagonal are 45°- 45° right triangles.

Diagonal *DB* forms
two right triangles
△*DAB* and △*DCB*

EXAMPLE 2 Finding the Diagonal of a Square

How long is the diagonal brace connecting the opposite corners on the back of a square frame that is 6 feet on a side?

SOLUTION

Make a sketch. Mark the right angle and the hypotenuse.

Since the length of each side of the 45° - 45° right triangles formed by the diagonal is 6 feet, the diagonal will be

$$6\sqrt{2} \approx 8.5$$

The length of the diagonal brace is approximately 8.5 feet.

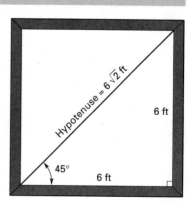

The measure of one angle of a right triangle is 45°. The length of one leg is 7.3 inches. How long is the other leg? How long is the hypotenuse?

The 30° - 60° Right Triangle

Another special right triangle has one angle equal to 30°. If one angle of a right triangle is 30°, how many degrees are in the other angle?

ACTIVITY 2   The Sides of a 30° - 60° Right Triangle

1 Use your protractor to draw three right triangles of different sizes, each having one angle of 30°. Label the 30° angle in each right triangle.

2 Label the shorter leg in each drawing with the letter *a*. Label the hypotenuse with the letter *c*. Notice that in each case the shorter leg of the triangle is opposite the 30° angle.

3 Complete the chart by carefully measuring the shorter leg *a* and the hypotenuse *c* in each of the triangles.

	a (side opposite the 30° angle)	*c* (hypotenuse)
Triangle 1	?	?
Triangle 2	?	?
Triangle 3	?	?

4 What is the ratio of *c* to *a* for each of the three triangles?

In the Activity, you found an important relationship. In a 30°- 60° right triangle, the hypotenuse is always twice the length of the shorter leg; that is,

$$c = 2a$$

To find the longer leg of a 30°- 60° right triangle, use the Pythagorean Theorem and the relationship $c = 2a$.

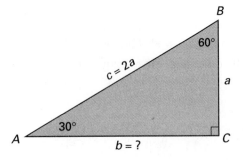

Use a sketch to label the sides and the angles. Label the side opposite the 30° angle a. Label the hypotenuse $2a$. Apply the Pythagorean Theorem.

$$a^2 + b^2 = c^2$$

Substitute $2a$ for c and solve for b.

$$a^2 + b^2 = (2a)^2$$

$$b^2 = 4a^2 - a^2$$

$$b^2 = 3a^2$$

Solving for b, the result is

$$b = \sqrt{3}a \text{ or } a\sqrt{3}$$

The leg AC (opposite the 60° angle) is equal to $\sqrt{3}$ times the length of the shorter leg. This is true for all 30°- 60° right triangles.

30°- 60° Right Triangle

For every 30°- 60° right triangle,

a. the hypotenuse is twice the length of the shorter leg.

b. the longer leg is $\sqrt{3}$ times the length of the shorter leg.

Ongoing Assessment

A 10-foot board is used as a ramp to slide crates from a warehouse floor to a loading platform. The board makes a 30° angle with the floor. What is the height (H) of the loading platform? What is the length of the horizontal distance (D) along the floor?

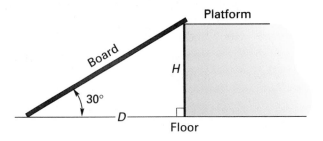

Important facts relating to the two special right triangles are summarized below.

45° – 45° Right Triangle

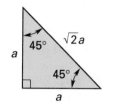

The legs have equal length. The length of the hypotenuse is $\sqrt{2}$ times the length of either leg.

3-4-5 Right Triangle

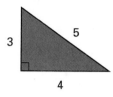

The 3-4-5 right triangle has sides whose ratios are 3:4:5.

30° – 60° Right Triangle

The shorter leg is opposite the 30° angle. The length of the hypotenuse is 2 times the length of the short leg. The length of the longer leg is $\sqrt{3}$ times the length of the short leg.

WORKPLACE COMMUNICATION

How many 3-4-5 right triangles are there in the satellite framework? How many pieces of each length of material should Raymond order?

Top View

Front View

45 cm

45 cm

|←— 60 cm —→|

NOTES ON DRAWING

1. Top and bottom faces are regular hexagons.
2. See SNAPSAT specification sheet for details of members and joinery.

CELSAT Engineering Drawing

Approved by:
Balram Gupta

MEMO FROM
Leigh Ann Fletcher
CELSAT Structural Analyst

TO: Raymond Villarreal
Structure Design & Analysis Technician

Raymond:

We are ready to fabricate a CELSAT structure framework for load and vibration testing. Please write a purchase order to Graphite Systems, Inc. to order the SNAPSAT frame members. Since these pieces are very expensive, only order the required number of each length and three replacement pieces for each length.

The CELSAT structure framework engineering drawing is attached. If you have any questions, please call me at ext 6-1059

Leigh Ann

Version:
FINAL

Series:
S-0193

LESSON ASSESSMENT

Think and Discuss

1 If a right triangle *ABC* is similar to a 3 - 4 - 5 right triangle, the hypotenuse of triangle *ABC* is a multiple of what number?

2 Why are 45°- 45° right triangles and 30°- 60° right triangles considered special right triangles?

3 How do you find the hypotenuse of a 45°- 45° right triangle when you know the length of a leg?

4 How do you find the hypotenuse of a 30°- 60° right triangle when you know the length of the shorter leg?

Practice and Problem Solving

5. Which of the following are right triangles?

a.   b.   c.   d.   e.

6. Find the length of *AC* if the hypotenuse is 12 inches.

7. Find the length of *BC* if the hypotenuse is 20 feet.

8. Find the length of *AB* if each leg is 8 centimeters.

9. The length of the hypotenuse of a 45°- 45° right triangle is 16 inches. Find the length of each leg.

M ∠A = 45°

10. Find the length of *XY* and the length of *XZ* if the length of *YZ* is 12 centimeters.

M ∠X = 60°

11. Find the length of *XZ* and the length of *YZ* if the length of *XY* is 6 feet.

12. Triangle *ABC* is an equilateral triangle. If the length of the altitude *AD* is $4\sqrt{3}$ inches, find the length of *BC*.

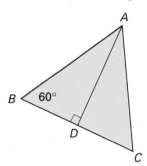

13. The length of the hypotenuse of a 30°- 60° right triangle is 20 inches. Find the length of each leg.

14. A bridge joins two points *A* and *B* across a river (see diagram). A surveyor determines that ∠*ABC* is 90° and ∠*ACB* is 60°. If the length of *BC* is 200 yards, how long is the bridge?

15. A carpenter is building trusses to support the roof of a new house. The trusses are each in the shape of a 3 - 4 - 5 right triangle. One leg of the triangle is 108 inches. The other leg is 144 inches. How long should the carpenter cut the lumber for the hypotenuse?

16. A 15-foot ladder is leaning against a building. The ladder makes a 30° angle with the building. How far from the building is the foot of the ladder?

17. Isaias is a catcher on his school's baseball team. He wants to practice throwing from home plate to second base with his dad. He knows the distance is the same as the diagonal of a square that has sides 90 feet long. How far apart should Isaias and his dad stand for Isaias to practice his throws?

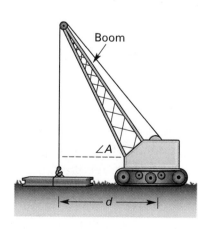

18. A crane with a 120-foot boom can safely lift a maximum weight of 10,000 pounds when the boom angle (∠*A*) is 60° or more. What is the maximum distance between the base of the crane and the weight when the crane lifts the maximum weight?

The following formula is used by a truck conversion business to find the number of four-wheelers (N) needed to break even. C is the start-up cost of production, S is the selling price of the four wheelers, and P is the cost of converting the four-wheelers.

$$N = \frac{C}{S - P}$$

19. Use the formula to find the number of four-wheelers that must be sold to break even if the start-up cost is $12,000, the selling price is $4000 per truck, and the production cost is $1000 per truck.

20. Solve the formula for the selling price (S).

Solve each equation for c.
21. 15% of $c = 180$ **22.** $c^2 + c = 20$

Write a system of equations to model each situation. Solve the system and check the solution.
23. Todd needs a total of 30 two-inch and three-inch bolts to assemble a piece of equipment. He needs six more two-inch bolts than three-inch bolts. How many two-inch bolts does Todd need?

24. Alicia needs 8 liters of a 55% saltwater solution for an experiment. She has one solution that is 40% saltwater and another that is 60% saltwater. How many liters of each solution should Alicia mix together to get a 55% saltwater solution?

Write a quadratic equation to model each situation. Solve the equation and check the solution.
25. The sum of the squares of two consecutive, even integers is 155 more than the odd integer between them. What are the two even integers?

26. A 9- by 12-foot rug is placed in a rectangular family room so that it leaves a uniform strip of flooring on all sides. The area of the exposed flooring is 100 square feet. What are the dimensions of the family room?

LESSON 12.5 TRIGONOMETRIC RATIOS

The hypotenuse of a right triangle is its longest side. It is also the side *opposite* the right angle. The other two sides of the right triangle are its legs.

In the drawing, the two legs are described in reference to angle *A*.

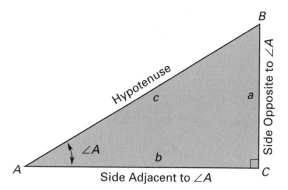

The Tangent Ratio

The tangent of an angle is an especially useful trigonometric ratio.

> **Tangent of an Angle**
> In a right triangle, the tangent of an acute angle is the ratio
> $$\frac{\text{Length of opposite leg}}{\text{Length of adjacent leg}}$$

The tangent of angle *A* is usually abbreviated

$$\tan A = \frac{\text{opp}}{\text{adj}}$$

Notice that the word *tangent* is shortened to *tan*. You can read the abbreviation as "tangent of angle A" or as "tan A." Using the labels in the drawing above,

$$\tan A = \frac{a}{b}$$

You can also write the tangent ratio for angle *B*

$$\tan B = \frac{b}{a}$$

Use the sketch of the 45°- 45° right triangle to answer these questions.

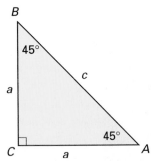

1 What is the tangent of ∠A and of ∠B?

2 What is the value of the tangent of ∠A if the legs of the 45°- 45° right triangle are both increased by the same amount? Explain your answer.

3 If an angle of a right triangle has a tangent of one, what can you conclude about the angles in the triangle?

You can find the slope of a line using the tangent ratio. Look at the drawing below. The slope is defined as the ratio of the rise to the run. The rise is the length of segment *BC*. The run is the length of segment *AC*.

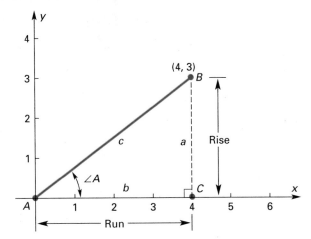

Thus, the slope of *AB* is

$$\text{Slope} = \frac{BC}{AC} = \frac{a}{b}$$

The ratio $\frac{a}{b}$ is also equal to the tangent of ∠A.

Slope and Tangent

The slope of a line is the same as the tangent of the angle formed by the line and the x-axis.

Critical Thinking   For a line that passes through the origin of the Cartesian coordinate system, there are two quadrants where the tangent of the angle formed by the line and the x-axis can be negative. What are the two quadrants?

EXAMPLE 1   Using the Tangent to Find the Slope

Find the slope of line segment AB.

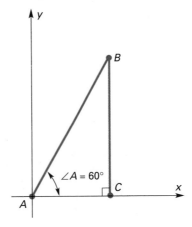

SOLUTION

Notice that $\angle A$ is opposite the leg of the right triangle that represents the rise of the line.

Now use a calculator to find tan A.

- Make sure your calculator is in the *degree mode*.
- Enter the measure of angle A.
- Press the [TAN] key.
- Read the value for the tangent of angle A.

Since m$\angle A$ is 60°, the tangent of $\angle A$ is about 1.7. The slope of the line is equal to the tangent of $\angle A$, which in this case is 1.7.

To check your result, measure the rise and run of the line and compute the slope using the slope formula. Is this ratio the same as the tangent of 60°?

The Sine and Cosine

The tangent of an angle is a useful ratio to use when you are solving problems involving right triangles. There are two other ratios that are also very useful.

> ### Sine and Cosine of an Angle
> In a right triangle, the sine of an acute angle is the ratio
>
> $$\frac{\text{Length of opposite leg}}{\text{Length of hypotenuse}}$$
>
> and the cosine of an acute angle is the ratio
>
> $$\frac{\text{Length of adjacent leg}}{\text{Length of hypotenuse}}$$

The sine and cosine of $\angle A$ are abbreviated like this:

$$\sin A = \frac{\text{opp}}{\text{hyp}} \qquad \cos A = \frac{\text{adj}}{\text{hyp}}$$

ACTIVITY 2   Trigonometric Ratios

1 Measure the lengths of the hypotenuse and legs of the right triangle.

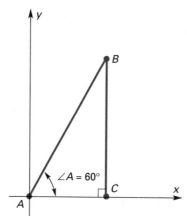

2 Use the lengths to calculate sin A and cos A.

3 Use your calculator to find sin B and cos B. Be sure your calculator is in degree mode before you enter the measure of $\angle A$.

4 How do the results from Step 2 and Step 3 compare?

Use the data from Activity 2 to show that the sine of an angle divided by its cosine is equal to its tangent.

EXAMPLE 2 Finding Trigonometric Ratios

The tangent of an angle in a right triangle is $\frac{5}{12}$. Find the sine of the angle.

SOLUTION

Draw a sketch.

The side opposite $\angle A$ equals 5. The side adjacent to $\angle A$ equals 12. The hypotenuse is c.

Use the Pythagorean Theorem or Pythagorean triple 5, 12, 13 to find that the hypotenuse c is equal to 13.

Thus, the sine of $\angle A$ is

$$\frac{\text{opp}}{\text{hyp}} \quad \text{or} \quad \frac{5}{13}$$

Ongoing Assessment

The cosine of an angle of a right triangle is $\frac{8}{17}$. Find the sine and tangent of the angle.

The branch of mathematics that uses the sine, cosine, and tangent ratios is called **trigonometry**. The word itself comes from *tri* (meaning "three") and *metry* (meaning "measure"). The sine, cosine, and tangent are called **trigonometric ratios**. Most early civilizations used trigonometry to measure distances indirectly. For example, people used the tangent ratio to find the angle between the horizontal and a line of sight toward a point above the horizontal. This angle is called the **angle of elevation**.

Angle of Elevation

Horizon

EXAMPLE 3 Using Indirect Measurement

The angle of elevation from a point on the ground to the top of a 35-foot cliff is 40°. How far is this point from the base of the cliff?

SOLUTION

To find the distance, draw a diagram and use the tangent ratio to write a trigonometric equation.

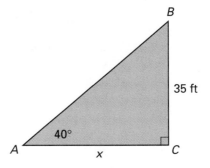

Solve for x.

$$x = \frac{35}{\tan 40}$$

$$x \approx \frac{35}{0.839} \text{ (Use your calculator to find } \tan 40° \approx 0.839)$$

$$x = 41.7$$

The distance to the base of the cliff is 41.7 feet (to the nearest tenth of a foot).

Ongoing Assessment

The angle of elevation from a point 300 feet from the base of a microwave relay tower to its top is 26°. What is the height of the tower?

LESSON ASSESSMENT

Think and Discuss

1 Compare the tangent ratio to the sine and cosine ratios. How are they alike? How are they different?

2 If you know the sine of one acute angle in a right triangle, how can you find the cosine of the other acute angle?

3 Explain how to use a calculator to find the sine of 60°.

4 Explain how to find the angle of elevation.

Practice and Problem Solving

Find the sine, cosine, and tangent for angle A and angle B. Write each ratio as a fraction in simplest form.

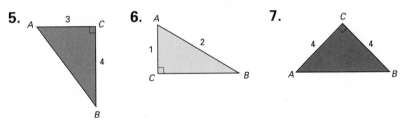

5.

6.

7.

Use a calculator to find the following. Round each answer to the nearest thousandth.

8. Sin 25°   **9.** Cos 45°   **10.** Tan 75°

Write the trigonometric equation that models each situation. Solve the equation, rounding to the nearest tenth.

11. You must cut a wire to connect the top of a 40-foot telephone pole with the ground. The wire must make a 35° angle with the ground. What length of wire should be cut?

40 ft

35°

12. From a point on the ground 80 feet from the base of a building to the top of the building, the angle of elevation is 30°. Find the height of the building.

13. What is the length of a shadow cast by a 40-foot flagpole when the angle of elevation of the sun is 40°?

The angle of depression is the angle between the horizontal and a line of sight below the horizontal. Write the trigonometric equation that models each situation. Solve each equation, rounding to the nearest tenth.

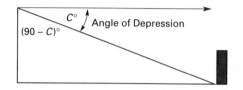

14. From a point on top of a 100-foot cliff, the angle of depression to a cabin below the cliff is 25°. How far is the cabin from the base of the cliff?

15. A lookout on a ship spots a school of whales. The lookout is standing 55 feet above the ocean surface and measures an angle of depression to the whales of 3°. How far from the ship is the school of whales?

16. A person standing on a 235-foot cliff spots a boat in a lake. The eye level of the person is 5 feet. The angle of depression to the boat is 30°. How far from the base of the cliff is the boat?

Mixed Review

Solve each equation for y.

17. $10(y + 3) = 8 + 6y$

18. $(y + 2)(y - 5) = 0$

Graph the system of inequalities. Shade the solution area.

19. $x + y \leq 1$
$2x - y > 2$

20. $x \geq 1$
$y \geq x$
$x + y < 5$

The function $P(a) = a^2 - 2a - 3$ is used to project profit for the Lite Shade Company.

21. Find the minimum value of P.

22. Find $P(0)$.

23. The center of a circle is located at $A(2, -3)$. One of the points on the circle is $B(5, 1)$. Use the distance formula to find the radius of the circle.

LESSON 12.6 SOLVING EQUATIONS CONTAINING RADICALS

Adding and Subtracting Radicals

800 ft

288 ft

A radical expression is used to find the time (in seconds) it takes an object to fall a given distance (in feet) when air resistance is not a factor.

Suppose a ball is dropped from a height of 800 feet. At the same time, another ball is dropped from 288 feet. How much time will elapse between each ball's impact with the ground? You can use the formula $t = \frac{1}{4}\sqrt{h}$ to find how many seconds it will take a ball to fall in h feet. The elapsed time (T) will be the difference in the time it takes each ball to reach the ground.

$$T = \frac{1}{4}\sqrt{800} - \frac{1}{4}\sqrt{288} \qquad \text{Given}$$

$$= \frac{1}{4}(20)\sqrt{2} - \frac{1}{4}(12)\sqrt{2} \qquad \text{Simplify radicals}$$

$$= 5\sqrt{2} - 3\sqrt{2} \qquad \text{Simplify fractions}$$

You can further simplify the difference $5\sqrt{2} - 3\sqrt{2}$. When an addition or subtraction expression contains terms with the same radicands, you can apply the Distributive Property.

$$= (5 - 3)\sqrt{2} \qquad \text{Distributive Property}$$

$$= 2\sqrt{2}$$

The time difference is about 2.8 seconds.

$5\sqrt{2}$ and $3\sqrt{2}$ are *like radicals* because they have the same radicands. You can add and subtract two like radicals in the same way you add and subtract other like terms. Thus,

$$5\sqrt{2} - 3\sqrt{2} = 2\sqrt{2}$$

Addition and Subtraction of Radicals

Expressions containing like radicals can be added and subtracted in the same way that like terms can be added and subtracted.

EXAMPLE 1 Adding Radical Expressions

Simplify $2\sqrt{3} + 2\sqrt{8} + 3\sqrt{12}$.

SOLUTION

Simplify the second and third radicals.

$$2\sqrt{3} + 2\sqrt{8} + 3\sqrt{12} = 2\sqrt{3} + 2\sqrt{4}\sqrt{2} + 3\sqrt{4}\sqrt{3}$$
$$= 2\sqrt{3} + 4\sqrt{2} + 6\sqrt{3}$$

Then, combine like radicals.

$$= 8\sqrt{3} + 4\sqrt{2}$$

Critical Thinking How can you use your calculator to show the simplification is correct?

Ongoing Assessment

Simplify $\sqrt{80} - \sqrt{125} + \sqrt{20}$.

Equations with Radicals

The following property is used to solve equations containing radicals.

Solving $\sqrt{x} = a$
If $\sqrt{x} = a$ and $x \geq 0$, then $x = a^2$.
That is, $(\sqrt{x})^2 = x$.

EXAMPLE 2 The Free Fall Formula

To calculate the energy water will produce in a hydroelectric plant, a technician needs to know the distance the water falls from the inlet of the dam to the turbine. The technician knows it takes the water 4 seconds to fall this distance. How can the technician find the distance?

SOLUTION

Solve the formula $t = \frac{1}{4}\sqrt{h}$ for h.

$$t = \frac{1}{4}\sqrt{h} \qquad \text{Given}$$

$$(t)^2 = \left(\frac{1}{4}\sqrt{h}\right)^2 \qquad \text{Square both sides}$$

$$t^2 = \frac{1}{16}h \qquad \left(\frac{1}{4}\right)^2 = \frac{1}{16} \text{ and } (\sqrt{h})^2 = h$$

$$h = 16t^2 \qquad \text{Multiplication Property of Equality}$$

When the time is 4 seconds, the distance the water falls is $h = 16(4)^2$, or 256 feet.

Critical Thinking   Explain why the following is true for all real numbers a and b: $(\sqrt{a} + \sqrt{b})(\sqrt{a} - \sqrt{b}) = a + b$

Sometimes an equation contains variables inside and outside the radical. To solve for a variable inside the radical:

- Isolate the radical.
- Square both sides of the equation.
- Solve the resulting equation.
- Check your solutions.

EXAMPLE 3 Radicals and Quadratic Equations

Solve the equation $\sqrt{3x - 2} = x$.

SOLUTION

$$\sqrt{3x - 2} = x \qquad \text{Given (The radicand is isolated.)}$$

$$(\sqrt{3x - 2})^2 = x^2 \qquad \text{Square both sides}$$

$$3x - 2 = x^2 \qquad (\sqrt{a})^2 = a$$

$$x^2 - 3x + 2 = 0 \qquad \text{Addition and Subtraction Properties of Equality}$$

$$(x - 2)(x - 1) = 0 \qquad \text{Factor the expression}$$

$$x = 2 \text{ or } x = 1 \qquad \text{Zero Product Property}$$

Thus, the solutions are 1 and 2. Check the solutions in the original equation.

If you square each side of an equation, the resulting equation is still balanced. However, the solutions may not make the original equation true.

For example, suppose $x = 4$. Then $x^2 = 16$. The solutions to $x^2 = 16$ are 4 and -4. However, -4 is not a solution to the original equation $x = 4$. In this case, -4 is an **extraneous solution.** The extraneous solution resulted from squaring both sides of the original equation.

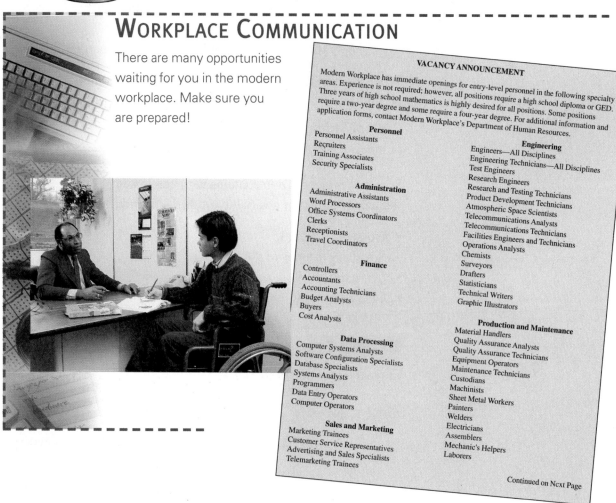

WORKPLACE COMMUNICATION

There are many opportunities waiting for you in the modern workplace. Make sure you are prepared!

VACANCY ANNOUNCEMENT

Modern Workplace has immediate openings for entry-level personnel in the following specialty areas. Experience is not required; however, all positions require a high school diploma or GED. Three years of high school mathematics is highly desired for all positions. Some positions require a two-year degree and some require a four-year degree. For additional information and application forms, contact Modern Workplace's Department of Human Resources.

Personnel
Personnel Assistants
Recruiters
Training Associates
Security Specialists

Administration
Administrative Assistants
Word Processors
Office Systems Coordinators
Clerks
Receptionists
Travel Coordinators

Finance
Controllers
Accountants
Accounting Technicians
Budget Analysts
Buyers
Cost Analysts

Data Processing
Computer Systems Analysts
Software Configuration Specialists
Database Specialists
Systems Analysts
Programmers
Data Entry Operators
Computer Operators

Sales and Marketing
Marketing Trainees
Customer Service Representatives
Advertising and Sales Specialists
Telemarketing Trainees

Engineering
Engineers—All Disciplines
Engineering Technicians—All Disciplines
Test Engineers
Research Engineers
Research and Testing Technicians
Product Development Technicians
Atmospheric Space Scientists
Telecommunications Analysts
Telecommunications Technicians
Facilities Engineers and Technicians
Operations Analysts
Chemists
Surveyors
Drafters
Statisticians
Technical Writers
Graphic Illustrators

Production and Maintenance
Material Handlers
Quality Assurance Analysts
Quality Assurance Technicians
Equipment Operators
Maintenance Technicians
Custodians
Machinists
Sheet Metal Workers
Painters
Welders
Electricians
Assemblers
Mechanic's Helpers
Laborers

Continued on Next Page

Now you are ready to complete the Math Lab Activities and Math Applications for this chapter.

MATH LAB

Equipment   Directional compass
Line level
Tape measure
Masking tape
Plumb line
Calculator
Cord or heavy string

Problem Statement

Construction contractors must grade the surfaces of parking lots to ensure adequate drainage of water. The amount of grade is related to the slope of the parking lot surface.

You will lay out a north-south line and an east-west line on a parking lot surface and measure the slopes of the lines. You will also calculate the angles corresponding to these slopes.

Procedure

a Stretch the cord taut and put two marks on the cord, 25 feet apart. These marks will be referred to as *A'* and *B'*.

b Choose a location in the parking lot at least 25 feet from any edge of the lot. Mark this location "point *A*" with a piece of masking tape.

c Use the compass to sight a line directly south of point A. Stretch the cord along this north-south line.

d Have two members of the group hold the ends of the cord at A' and B'. Make sure the cord is taut. One member should stand at point A and hold the cord at a comfortable height. Use the plumb line as shown to locate A' directly above point A. Measure the vertical distance along the plumb line from point A to A'. Label this distance H_A and record its value. The group member holding the cord at point A should keep the vertical distance H_A constant and keep the A' mark directly above point A throughout the remainder of this Activity.

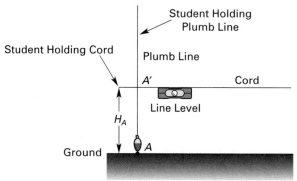

e Attach the line level to the cord near the B' mark. Adjust the height of the cord until the line level indicates the cord is level.

f After leveling the cord, use the plumb line to locate the point on the ground directly beneath the B' mark. Mark this location "point B" with a piece of masking tape. Measure the vertical distance from point B to B'. Label this distance H_B and record its value.

g Return to point A. Use the directional compass to sight a line directly west of point A. Stretch the cord along this east-west line.

h Repeat Steps **d** through **f** to locate point C and to measure the height of the level cord above point C.

i Use the formula

$$\text{Slope} = \frac{\text{Rise}}{\text{Run}}$$

where run is the length of the cord between A' and B', to calculate the slope of the north-south line. In this case

$$\text{Run} = 25 \text{ ft}$$

$$\text{Rise} = H_A - H_B$$

so that

$$\text{Slope} = \frac{H_A - H_B}{25}$$

The drawing below shows the geometry used to calculate the slope.

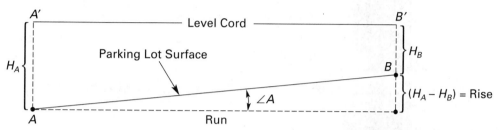

If H_A is greater than H_B, the slope of the north-south line from A to B is uphill, and the slope is positive. If H_A is less than H_B, the slope of the north-south line from A to B is downhill, and the slope is negative. Record the slope of the north-south line.

j Use the same formula to calculate the slope of the east-west line. In this case,

$$\text{Run} = 25 \text{ ft}$$

$$\text{Rise} = H_A - H_C$$

and

$$\text{Slope} = \frac{H_A - H_C}{25}$$

If H_A is greater than H_C, the slope of the east-west line from A to C is uphill, and the slope is positive. If H_A is less than H_C, the slope of the east-west line from A to C is downhill, and the slope is negative. Record the slope of the east-west line.

k Now calculate the angles corresponding to the slopes you just measured. Look at the drawing again. Find m∠A from the tangent function. Suppose you measure the rise $(H_A - H_B)$ to be 0.5 feet.

$$\tan A = \frac{\text{opp}}{\text{adj}} = \frac{0.5}{25} = 0.02$$

To find m∠A, do the following:

- Place your calculator in the degree mode.
- Enter the value 0.02.
- Press the (INV) inverse or 2nd function key.
- Press the (TAN) key.
- Read ∠A (given in degrees) in the window. For tan A = 0.02, ∠A (rounded) is 1.1°.

Calculate the measures of the two angles—one for the right triangle whose hypotenuse is segment AB and the other whose hypotenuse is AC. A positive angle means the parking surface runs uphill from A to C or from A to B. A negative angle means the surface runs downhill. Do you expect good drainage from the parking lot? Explain your answer.

Activity 2: Calculating Diagonal Lengths of Rectangles

Equipment   Tape measure
Calculator

Problem Statement

Drawing a diagonal across a rectangle results in two identical right triangles.

You will measure the length and the width of several rectangular objects. Then, you will calculate the lengths of the diagonals using the Pythagorean Theorem and compare the calculated values to the measured values. In addition, you will calculate the length of a room diagonal from a corner at the ceiling to an opposite corner at the floor.

Procedure

Complete Steps **a**, **b**, and **c** for each of the following rectangular objects:

- classroom door
- teacher's desk
- chalkboard or dry erase board
- classroom floor

a Measure and record the width (w), length (l), and diagonal (D).

b Use the Pythagorean Theorem to calculate the length of the diagonal.

c Compare the calculated diagonal lengths to the measured diagonal lengths.

d Measure the height (h) of the classroom. Use the length, width, and height to find the diagonal of the room from one corner at the ceiling to the opposite corner at the floor. Find the diagonal using the formula

$$D^2 = l^2 + w^2 + h^2$$

e How does the length of the diagonal of the room compare with the length, width, and height of the room?

Activity 3: Calculating the Length of One Side of a School Building

Equipment   Directional compass
Flat square
String (100 feet)
Calculator

Problem Statement

You will use the tangent ratio to estimate the length of one side of your school building.

Procedure

a Choose a corner of the school building (see drawing). Extend the 100-foot string from the corner of the school building at a right angle to the side being measured. Use the flat square to determine a 90° angle.

b From the end of the string farthest from the building, use the directional compass to find the heading to the chosen corner. Find a second compass heading from the end of the string to the opposite corner of the side of the building being measured (see drawing on next page). Record the two compass headings.

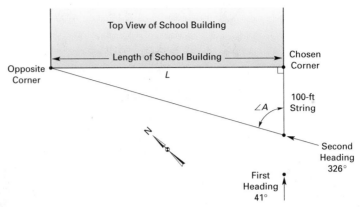

c Subtract the two compass headings. If the difference between the compass headings is less than 90°, this difference is the measure of ∠A in the drawing. If the difference is greater than 90°, subtract the difference from 360° to get m∠A.

For example, suppose the compass headings were 41° and 326°, as shown in the drawing. This gives 326° − 41° = 285°. Since 285° is greater than 90°, subtract 285° from 360° to get 75°. The measure of ∠A is 75°.

d Use the formula below to calculate the distance between the two corners.

$$\tan A = \frac{\text{opp}}{\text{adj}} = \frac{\text{length of side of school building}}{\text{Length of string}}$$

$$\tan A = \frac{L}{100 \text{ feet}}$$

$$L = 100 \tan A$$

e Describe how you could use this method to measure the distance between two points on opposite sides of the Grand Canyon.

Two streets intersect at an angle of 60°. The perpendicular distance across each street is 35 feet. At the intersection, the distance (*AB*) from corner to corner is greater than 35 feet.

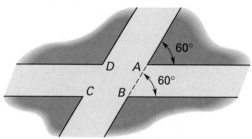

1 Trace the drawing of the intersection onto your own paper. Draw a right triangle using the distance between two corners as the hypotenuse and the perpendicular distance across one of the streets as one of the legs. Label the measures of the angles within your right triangle.

2 Determine the distance between the two street corners *A* and *B*.

3 What is the distance between corners *C* and *D*?

4 You are building a square picture frame from a kit. Each of the four frame pieces is 12 inches long. To check for "squareness," you measure the diagonals to see if they are equal. Find the length of the diagonals connecting the outside corners of the square frame.

5 A trail map shows that you are 20 miles away from a mountain peak. You can see the peak in the distance, and you estimate that it is about 7° above the horizon. Draw a sketch to illustrate this information. Use the tangent ratio to estimate the height (to the nearest 1000 feet) of the mountain peak.

6 A single-engine airplane loses power at an altitude of 6000 feet. The plane can maintain a glide path that makes an angle of depression of at least 18°. What is the maximum horizontal distance the plane can glide before it makes an emergency landing?

7 A television antenna is mounted on a mast at the peak of a roof. When standing 40 feet away from the end of the building, you measure the angle of elevation to the top of the antenna as 50° and the angle to the peak of the roof as 39°. Determine the height of the peak of the roof, the height of the antenna above the ground, and the length of the mast.

A research team reflects a laser beam from a mirror placed on the moon by the Apollo astronauts. The laser beam has a very small angle of spread, about 0.001°. The beam travels 239,000 miles to reach the moon.

8 Draw two right triangles representing the beam as it spreads from its source toward the moon. Label the angle of spread (known as the *divergence angle*), each triangle, and the length of the leg that corresponds to the distance to the moon. (*Hint:* Half the beam is used for each right triangle.)

9 Determine the total width of the beam when it reaches the moon.

You measure the angle that the sun makes with the vertical by measuring the shadow cast by a perpendicular object of known height. Suppose a yardstick that is perpendicular to the ground casts a shadow that extends 9 inches from the base of the yardstick.

10 Make a sketch showing the right triangle's dimensions and the angle that the sunlight makes with the yardstick.

11 What is the value of the tangent of the angle formed by the sunlight with the yardstick?

12 When the tangent of angle *A* is known, the inverse tangent function will give the value of angle *A*. Enter the tangent value into your calculator and press the (INV) key followed by the (TAN) key. The calculator will then display the angle. What is the angle of the sun in the measurement above?

Astronomers measure large distances using an angular measurement called *parallax*. The position of a distant object is precisely charted. Six months later, when the Earth is on the opposite side of the sun, the object is charted again. The position of the object in the sky will be at a slightly different angle. The drawing below illustrates this for a certain star. (*Note:* the drawing is not to scale.)

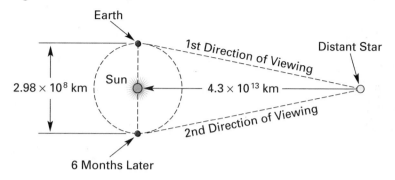

13 Identify the right triangles in the drawing. Note that the small angle in each right triangle is $\frac{1}{2}$ the angle determined from the two chartings. Identify the adjacent and opposite legs with respect to this small angle.

14 Compute the tangent of the small angle in one of the right triangles.

15 Use the INV key followed by the TAN key (see Exercise 12) to find the measure of the small angle in the right triangle.

16 Double the measure of the small angle. This angle is referred to as *stellar parallax*. The distance used in this Exercise is for the closest star to the sun and involves a relatively large amount of stellar parallax. Astronomers use much smaller angles in actual practice.

Pendulum clocks can maintain accurate time because they have a finely adjustable period of swing. The period of swing of a pendulum is given by the formula

$$T = 2\pi \cdot \sqrt{\frac{L}{g}}$$

T is the period of the pendulum in seconds (the period is defined as the time required for one complete swing, back and forth).
L is the length of the pendulum in centimeters.
g is the acceleration due to gravity, 980 centimeters per second.

17 To what length should you adjust the pendulum for a 1.000 second period? Use your calculator's value for π. Write your answer to the nearest thousandth.

18 If the length is shortened to 24.800 centimeters, would the period be longer or shorter than 1 second?

While repairing a fence, you need to determine whether the fence's corner forms a 90° angle. Both sections of fence are straight and have poles spaced 8 feet apart. Your only measuring device is a 50-foot tape measure.

19 Describe a procedure incorporating a Pythagorean triple and a 3-4-5 right triangle that will determine whether the fence lines are perpendicular.

20 You are using the Doyle log rule to estimate the amount of wood to haul on a log truck. To use the rule, you measure the diameter of the small end of a log (in inches) and its length (in feet), then use the following formula:

$$BF = L \cdot \left(\frac{d-4}{4}\right)^2$$

BF is the approximate board feet of usable lumber.
d is the log's small-end diameter in inches.
L is the length of the log in feet.

For cost effectiveness, the truck must haul about 5000 board feet of usable lumber. If the truck can carry 20 logs, each about 27 feet long, estimate the small-end diameter of the logs for a 5000 board-foot load. Write your answer to the nearest inch.

When a business sends out a mass-mail advertisement, it must consider the expected response rate and its confidence interval. The expected response rate is the expected increase in sales (expressed as a percent) resulting from the advertisement. The confidence interval is specified as the upper and lower limit of the sales increase. For example, a 95% confidence interval means that 95 mass mailings out of 100 will result in at least the lower limit specified and at most the upper limit specified.

The limits for a 95% confidence interval can be calculated from the formulas below:

$$L_U = p + 1.96 \cdot \sqrt{\left(\frac{p \cdot (1-p)}{n}\right)}$$

$$L_L = p - 1.96 \cdot \sqrt{\left(\frac{p \cdot (1-p)}{n}\right)}$$

L_U is the upper limit of the response (95% of the time).
L_L is the lower limit of the response (95% of the time).
p is the response rate (a decimal).
n is the number of pieces mailed.

21 A business wants to use an advertisement to achieve a 3% response rate and a 95% confidence interval of 1.34%. Assume half the confidence interval is above the expected response rate and half is below. This means that

$$L_L = 0.03 - \frac{0.0134}{2} = 0.0233$$

How many pieces must be mailed to achieve this lower limit?

22 Substitute this value into the formula for L_U and solve for L_U. Subtract L_L from L_U. Does your answer agree with the desired 95% confidence interval?

HEALTH
OCCUPATIONS

An X-ray source is positioned over a sheet of film, as shown below. The source is centered 40 inches away from the 10-inch-wide film sheet.

23 How far must the X-rays travel to reach the edges of the film?

24 Proper exposure of the film is very sensitive to the intensity of X-rays. The intensity (I) of the X-rays follows the inverse square law. If I_O is the X-ray intensity at a distance d_O, then the intensity at distance d is given by

$$I = \frac{I_O d_O^2}{d^2}$$

What is I at the center of the film? What is I at the edge of the film? Write each intensity as a function of $I_O d_O^2$.

25 How much more intense are the X-rays at the film's center than at the film's edges? (Write the ratio of the X-ray intensity at the edge of the film to the intensity at the center of the film.)

26 When taking a series of X-rays of the lumbar spine, one view requires the patient to sit and lean against a foam support at a 45° angle. If the distance from the base of the spine to the neck is 66 centimeters, to what height must the neck be raised to create a 45° angle? (Assume the spine is straight.)

Another X-ray view requires the X-ray source to be positioned at an angle of 20° from the centerline and to be located 40 inches vertically from the table.

27 What is the horizontal distance of the X-ray source from the centerline?

28 Will the X-ray source be more than or less than 40 inches from the center of the film? How much more or less will it be than 40 inches?

Nuclear medicine technicians caring for patients with radioactive implants must minimize their own exposure to radiation. One of the factors controlling the amount of radiation health-care workers receive is distance from the patient.

The radiation level follows the *inverse-square law*; a certain radioactive implant produces a radiation intensity I at a distance z given by

$$I = 100 \text{ mrems per hour} \cdot \frac{z^2_{\,O}}{z^2}$$

where $z_O = 1$ meter.

29 At what distance is the radiation intensity 50 mrems per hour? 10 mrems per hour? 1 mrem per hour?

FAMILY AND CONSUMER SCIENCE

An art display includes two sound effects speakers. You want to position the speakers on the wall so that when you stand midway between them, you are 7 feet from each speaker. Both speakers should be the same distance from the corner.

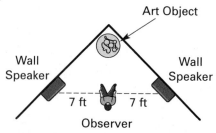

30 How far from the corner should you position each speaker?

INDUSTRIAL TECHNOLOGY

You are constructing a form for a concrete sidewalk that makes a 90° angle with the curb. Before you pour the concrete, you must verify that the forms are *square*; that is, whether they form 90° angles at the corners. The forms are 12 feet long and 4 feet wide.

Curb — 12 ft — Porch — 4 ft

31 Describe a procedure, using the Pythagorean Theorem, your tape measure, and a calculator, which verifies whether your forms are *square*.

A guy wire is used to stabilize a utility pole. The wire is fastened to the pole 25 feet above the ground and staked to the ground 4 feet from the base of the pole. Someone recommends that the pole is more stable if the ground stake is moved out to 6 feet from the pole.

32 How long is the guy wire when it is staked 4 feet from the base of the pole?

33 How long must the guy wire be if it is staked 6 feet from the pole?

In electrical circuits with varying currents and voltages, the combined effect of resistance and reactance is called impedance. The impedance is related to the resistance and the reactance by the following formula:

$$Z^2 = R^2 + X^2$$

Z is the impedance in ohms.
R is the resistance in ohms.
X is the reactance in ohms.

34 Draw and label a right triangle to represent the relation between the impedance, the resistance, and the reactance.

35 A stereo speaker is labeled as having an impedance of 8 ohms. Your measurement of the speaker circuit with an ohmmeter shows a resistance of 1.5 ohms. What is the reactance of the speaker?

A safety manual illustrates guidelines for the safe use of an extension ladder. Suppose your ladder has a maximum extended length of 24 feet.

36 What is the maximum distance *A* to the top support for your ladder?

37 What is the maximum landing height that you can safely reach with your ladder?

A plane is forced to travel the two-legged path shown below to avoid flying over a restricted area. Point *B* is 3.4 kilometers from a direct route through the restricted area.

38 How much shorter is the direct route than the two-legged path?

You must load material onto a platform that is 26 inches above the ground. You want to use a ramp, but the ramp angle must not be greater than 30°.

39 Draw a triangle that represents this situation. Determine the shortest ramp you can use.

40 Using a ramp of this length, what is the horizontal distance from the platform to the point where the ramp touches the ground?

41 A surveyor made an angular error of 0.05° when surveying a rectangular lot that is 350 feet wide. What is the error in the width of the lot?

You are designing a gardening shed with a sloped roof.

42 Write a formula for the tangent of the roof angle using the rise and the run for the roof.

43 Substitute a run of 12 feet into your formula and isolate the variable for the rise. Use this formula to make a table of the rise for roof angles of 5°, 10°, 15°, 20°, 25°, and 30°.

44 For each angle, compute the length of roof decking needed to span the distance between the two walls. This will be the hypotenuse of the right triangle. Add these values to your table.

45 If the run remains constant at 12 feet, what are some of the consequences of using too large an angle for the roof?

A certain design of a screw has 10 thread grooves in 1.5 inches. Thus, the width of the groove is 0.15 inches. The spread of the groove is 60°. The groove width and spread are shown in the drawing below.

46 Draw and label a right triangle that shows the known angles and dimensions for a thread groove.

47 What is the depth *d* of this thread?

You are asked to help design a cylindrical tank with a diameter of 24 feet. The tank will have a cone-shaped roof rising 5 feet above the lip of the tank. It will be made of steel weighing 40 pounds per square foot.

48 What is the slant height of the roof's surface; that is, the distance from the peak to the rim of the cylindrical tank?

49 What is the surface area of the roof? (*Hint:* The surface area of the cone is given by the formula $A = \frac{1}{2}\pi sd$, where s is the slant height and d is the diameter.)

50 Multiply the surface area (in square feet) by the weight of the steel per square foot to find the estimated weight of the tank's steel roof.

To estimate the height of a tall broadcasting antenna tower, you drive exactly $\frac{1}{2}$ mile from the tower's base. Using your protractor, you estimate the top of the tower to be 20° above the horizon.

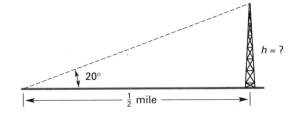

51 Write the trigonometric ratio that relates the angle to the dimensions of a right triangle.

52 Find the height of the tower.

53 What are some possible sources of error in this procedure?

A jet maintains a speed of 600 kilometers per hour while climbing at an angle of 30°.

54 You can represent the velocity of the jet using two vectors. One vector shows the horizontal velocity, and the other vector shows the vertical velocity. Draw a right triangle that shows the horizontal and vertical velocity components, as well as the resultant velocity of the jet at a 30° angle. The resultant velocity vector is the hypotenuse.

55 Determine the value of the horizontal and vertical components of the jet's velocity. (*Hint:* Find the length of the legs of the triangle you drew in Exercise 48.)

56 Use the vertical component of the velocity to find how long it will take the jet to climb 10 kilometers.

Mechanical drawing sets have triangles of various sizes. Each dimension in the chart below is the length of the longest leg. For each triangle, determine the length of the other leg and the length of the hypotenuse.

30° - 60° - 90° Drawing Triangles

57 4 in.

58 6 in.

59 10 in.

60 18 in.

61 24 in.

45° - 45° - 90° Drawing Triangles

62 4 in.

63 6 in.

64 10 in.

65 18 in.

66 24 in.

You are installing a television antenna on a mast on the roof of a house. You must determine the lengths of the guy wires that will extend from the mast to the eaves on the corners of the house. The wires will connect 15 feet above the base of the mast. The mast will rest atop the house peak, which is 7 feet above the level of the eaves. The eaves form a square that is 40 feet on a side, and the peak is centered in the square.

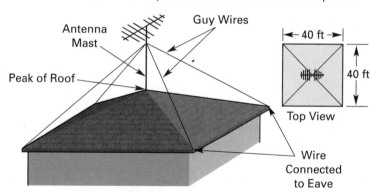

67 Determine the length of guy wire needed between the mast and each corner.

An automotive windshield repair technician is checking the specifications of a replacement for a car's rear window. He must determine the angle that the car's rear window makes with the window ledge. The depth of the ledge is 12 inches, and its corresponding height is $10\frac{1}{2}$ inches.

68 Make a sketch of the right triangle formed by the rear window, the window ledge, and the height. Label the known dimensions. Identify the slope of the window surface.

69 Compute the tangent of this angle.

70 To find the angle, enter the value of the tangent (that is, the ratio) into your calculator. Press the (INV) key followed by the (TAN) key. What angle should appear in the rear window specifications?

A home buyer has asked the builder to change the design of a house roof. The current plans call for the roof to have a rise-to-run ratio of 1:3, using 16-foot rafters with an 18-inch overhang. The buyer has asked to change the slope to a ratio of 2:5.

71 What is the tangent of the angle formed by the rafter and the plate for the roof slope of the current design?

72 To find the angle, enter the value of tan *A* into your calculator, and press the (INV) key followed by the (TAN) key. What angle do the rafters make with the plate for the current design and for the new design?

73 Is the new slope steeper or flatter than the current design slope?

74 The builder has already bought materials. He must still use 16-foot rafters and keep the same run of 13.75 feet. With the new roof slope, will the amount of overhang be more or less than the original overhang?

75 What is the length of the new overhang?

Fire fighters compute the rate of flow of water through a fire hose with the following formula:

$$\text{Flow Rate} = 30 \cdot d^2 \cdot \sqrt{p}$$

Flow rate is in gallons per minute.
d is the nozzle diameter in inches.
p is the nozzle pressure in pounds per square inch (psi).

76 A fire chief is designing a training exercise for his station. He wants the crew to use two hoses with a combined flow rate of 1000 gallons per minute. The nozzle diameter of each hose is 2.5 inches, but the nozzle pressure of one hose is four times the nozzle pressure of the other hose. What nozzle pressure should the chief use for the hoses?

CHAPTER 12 ASSESSMENT

Skills

Solve each equation.

1. $m^2 + 144 = 169$

2. $\sqrt{6} \cdot \sqrt{12} = t$

3. $q = \dfrac{\sqrt{75}}{\sqrt{15}}$

4. $\sqrt{2d - 3} = 5$

Applications

5. Triangle ABC and triangle XYZ are similar right triangles. Suppose the length of AC is 15 meters, BC is 9 meters, and XZ is 35 meters. What is the length of XY?

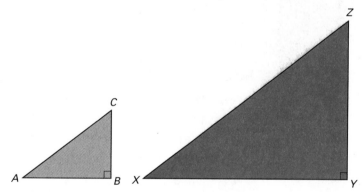

6. One leg of a right triangle is 8 inches. The other leg is 15 inches. How long is the hypotenuse?

7. The hypotenuse of a right triangle is 13 feet. One leg is 5 feet. How long is the other leg?

8. How long is the diagonal of a square that is 3 inches on each side?

9. The legs of a 45°- 45° right triangle are 12 millimeters each. What is the length of the hypotenuse?

10. The hypotenuse of a 30°-60° right triangle is 9 centimeters. How long is each leg?

Write each ratio as a fraction in lowest terms.

11. Sine of angle *A*.

12. Cosine of angle *A*.

13. Tangent of angle *B*.

14. The angle of elevation to the top of a 125-foot building is 50°. How far from the base of the building was the sighting taken?

15. A ramp is 150 meters long. It rises vertically for a distance of 30 meters. What is the angle of elevation for the ramp?

Math Lab

16. In the drawing below, the cord is held level between *A'* and *B'* 25 feet apart. The height measurement from *A* to *A'* is 3 feet 5 inches. The height measurement from *B* to *B'* is 2 feet 11 inches. Determine the slope of the parking lot.

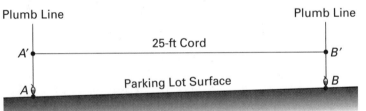

17. The dimensions of a classroom are 10 meters by 6.1 meters by 3 meters. What is the length of a diagonal that runs from the right front ceiling corner of the classroom to the left rear corner of the classroom?

18. Angle *A* is a right angle. The distance from point *A* to point *C* is 50 meters. The two compass headings at point *C* are 213° and 155°. How far is it from point *A* to point *B*?

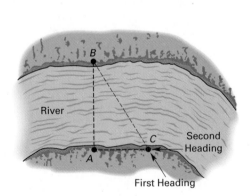

Tables of Conversion Factors

To convert from meters to inches, for example, find the row labeled "1 meter" and the column labeled "in." The conversion factor is 39.37. Thus, 1 meter = 39.37 in.

LENGTH

	cm	m	km	in.	ft	yd	mi
1 centimeter	1	0.01	10^{-5}	0.3937	3.281×10^{-2}	1.094×10^{-2}	6.214×10^{-6}
1 meter	100	1	10^{-3}	39.37	3.281	1.094	6.214×10^{-4}
1 kilometer	10^5	1000	1	3.937×10^4	3281	1094	0.6214
1 inch	2.54	0.0254	2.54×10^{-5}	1	0.0833	0.0278	1.578×10^{-5}
1 foot	30.48	0.3.48	3.048×10^{-4}	12	1	0.3333	1.894×10^{-4}
1 yard	91.44	0.9144	9.144×10^{-4}	36	3	1	5.682×10^{-4}
1 mile	1.6093×10^5	1609.3	1.6093	6.336×10^4	5280	1760	1

AREA

	cm²	m²	in.²	ft²	acre	mi²	ha
1 square centimeter	1	10^{-4}	0.1550	1.076×10^{-3}	2.471×10^{-8}	3.861×10^{-11}	10^{-8}
1 square meter	10^4	1	1550	10.76	2.471×10^4	3.861×10^{-7}	10^{-4}
1 square inch	6.452	6.452×10^{-4}	1	6.944×10^{-3}	1.594×10^{-7}	2.491×10^{-10}	6.452×10^{-8}
1 square foot	929.0	0.09290	144	1	2.296×10^{-5}	3.587×10^{-8}	9.29×10^{-6}
1 acre	4.047×10^7	4047	6.273×10^6	43,560	640	1.563×10^{-3}	0.4047
1 square mile	2.590×10^{10}	2.590×10^6	4.007×10^9	2.788×10^7	640	1	259
1 hectare	10^8	10^4	1.55×10^7	1.076×10^5	2.471	3.861×10^{-3}	1

VOLUME (CAPACITY)

	cm³	m³	in.³	ft³	L	oz	gal
1 cubic centimeter	1	10^{-6}	0.06102	3.531×10^{-5}	1.000×10^{-3}	0.03381	2.642×10^{-4}
1 cubic meter	10^6	1	6.102×10^4	35.31	1000	3.381×10^4	264.2
1 cubic inch	16.39	1.639×10^{-5}	1	5.787×10^{-4}	0.01639	0.5541	4.329×10^{-3}
1 cubic foot	2.832×10^4	0.02832	1728	1	28.32	957.5	7.480
1 liter	1000	1.000×10^{-3}	61.03	0.03532	1	33.81	0.2642
1 ounce	29.57	2.957×10^{-5}	1.805	1.044×10^{-3}	0.02957	1	7.813×10^{-3}
1 gallon	3785	3.785×10^{-3}	231	0.1337	3.785	128	1

1 gallon = 4 quarts (qt) = 8 pints (pt) = 16 cups (c)
1 cup (c) = 8 ounces (oz) = 16 tablespoons (tbsp) = 48 teaspoons (tsp)

MASS/WEIGHT

	g	kg	oz	lb	ton (short)	ton (metric)
1 gram	1	10^{-3}	0.03527	2.205×10^{-3}	1.102×10^{-6}	10^{-6}
1 kilogram	10^3	1	35.27	2.205	1.102×10^{-3}	10^{-3}
1 ounce	28.35	0.02835	1	0.0625	3.125×10^{-5}	2.835×10^{-5}
1 pound	453.6	0.4536	16	1	0.0005	4.536×10^{-4}
1 ton (short)	9.072×10^5	907.2	3.2×10^4	2000	1	0.9072
1 ton (metric)	10^6	10^3	3.527×10^4	2.205×10^3	1.102	1

ANGLE

	min (')	deg(°)	rad	rev
1 minute	1	0.01667	2.909×10^{-4}	4.630×10^{-5}
1 degree	60	1	0.01745	2.778×10^{-3}
1 radian	3438	57.30	1	0.1592
1 revolution	2.16×10^4	360	6.283	1

1 gallon = 4 quarts (qt) = 8 pints (pt) = 16 cups (c)

1 cup (c) = 8 ounces (oz) = 16 tablespoons (tbsp) = 48 teaspoons (tsp)

TIME

	s	min	hr	d*	y*
1 second	1	0.01667	2.788×10^{-4}	1.157×10^{-5}	3.169×10^{-8}
1 minute	60	1	0.01667	6.944×10^{-4}	1.901×10^{-6}
1 hour	3600	60	1	0.04167	1.141×10^{-4}
1 day*	8.640×10^4	1440	24	1	2.738×10^{-3}
1 year*	3.156×10^7	5.259×10^5	8766	365.3	1

Glossary

Absolute value The distance of a number from zero on a number line.

Absolute value of a vector The length of a vector, its numerical value. Also called the magnitude.

Addition property of equality When you add the same quantity to both sides of an equation, the equation stays equal and balanced.

Addition property of inequality When you add the same quantity to both sides of an inequality, the inequality stays equal and balanced.

Additive identity Zero is called the additive identity because when you add zero the value of the original quantity remains unchanged.
Example: $5 + 0 = 5$

Algebraic expression A mathematical sentence involving at least one variable.

And An operator that connects two mathematical sentences and indicates that all solutions must satisfy *both* sentences. The system would be a conjunction.

Angle of elevation The angle between a horizontal line (parallel to the ground) and a line drawn from a given point to some point above the ground (usually the top of a tall object).

Area The amount of space inside a flat shape, such as a triangle or a circle.

Array A rectangular arrangement of numbers.

Associative property of addition When you add multiple terms, you can change the way they are grouped without changing the answer.
Example: $5 + (3 + 17) = (5 + 3) + 17$

Associative property of multiplication When you multiply several terms, you can change the way they are grouped without changing the answer.
Example: $5 \times (3 \times 17) = (5 \times 3) \times 17$

Axis A number line used to locate points on a graph. Locating points on a plane requires two axes.

Axis of symmetry A line drawn through the vertex of a parabola such that the part of the parabola on one side of the line is the exact mirror image of the part on the other side.

Balancing equations Whenever you perform an operation on one side of an equation, you must perform the same operation to the other side in order for the equation to remain true.

Base Whenever a number is raised to a power, that number is called the base.
Example: in $2^3 = 8$, 2 is the base

Basic Property An algebraic expression that is true for every possible value of the variables it contains.

Binomial A polynomial with two terms.

Binomial square The product of two identical binomials. A binomial square produces a perfect square trinomial.

Break-even point The point at which two equations have the same value.

Cartesian coordinate system A method of locating points on a plane. The system uses two axes that cross at right angles to each other. The axes are marked off in units starting at zero where the axes cross. One axis runs vertically, and its numbers increase as you move up the axis—this is called the y-axis. The other axis—the x-axis—runs horizontally, and its numbers increase as you move to the right.

Circumference The distance around a circle, starting at one point, going around the circle once and ending up at the same point you started.

Closed interval An interval that includes its endpoints.

Coefficient When a variable is multiplied by a constant, that constant is called a coefficient of the variable. Example: in the expression $4x - 2y = 12$, 4 and 2 are coefficients

Combining vectors Given two vectors A and B, the process of placing the tail of vector A at the head of vector B and then drawing a new vector starting at the tail of vector B and ending at the head of vector A. This new vector is called the resultant vector.

Common denominator The least common multiple of all the denominators in a problem. A common denominator allows you to combine fractions easily.

Common factors When you list all the factors of two or more quantities, the common factors are those items that appear in all of the lists.

Commutative property of addition When you add several terms, you can change the order of addition without affecting the final answer. Example: $23 + 6 = 6 + 23$

Commutative property of multiplication When you multiply several terms, you can change the order of multiplication without affecting the final answer. Example: $23 \times 6 = 6 \times 23$

Completing the square A process of factoring a polynomial by adding or subtracting a constant in order to create a perfect square trinomial.

Composite number A number that has factors other than one and the number itself; that is, it has more than one factor pair.

Compound inequality A system of more than one inequality joined by AND or OR operators.

Compound interest Interest computed using both the principal and any previously accrued interest.

Conditional inequality An inequality whose truth depends on the value of a variable, such as $x > 5$.

Congruent Two figures are congruent if they are the same point for point.

Conjunction A compound inequality involving the AND operator.

Consistent system A system of equations that has at least one solution.

Constant A quantity whose value does not change, such as 5 or 12. Pi is also a constant since it always equals 3.14, rounded to the nearest hundredth.

Constant term The term in a polynomial that has no variable part.

Constraint Limits on the possible solutions to a problem. In a linear programming problem, each inequality is called a constraint.

Coordinate A number that is used to tell where a point is located on a graph. The ordered pair (7, 2) gives two coordinates that locate a unique point in the Cartesian coordinate system.

Correlation coefficient A number between -1 and 1 that tells how two sets of data are related. A value near 1 or -1 shows high correlation (positive or negative, respectively), while a value near zero shows a lack of correlation.

Corresponding angles In similar triangles, the angles that are equal between the two triangles are called corresponding angles.

Corresponding sides In similar triangles, the corresponding sides are the sides opposite the corresponding angles.

Cosine The ratio of the adjacent side of an angle to the hypotenuse in a right triangle.

Counterexample A set of values that makes an equation false. One counterexample is all that is needed to prove an equation is not a basic property.

Cross-multiplying Multiplying the means and extremes of a proportion. If $\frac{3x}{4} = \frac{5}{12}$, then the result of cross-multiplying is $4(5) = 3x(12)$.

Cube of number The third power of a number. Example: 5^3 is five cubed, or 125.

Cube root If $x^3 = y$, then x is a cube root of y.

Density The property that between any two real numbers there is another real number; the rational numbers are also dense, the integers are not.

Dependent events When the outcome of one event affects the possible outcomes of another, the events are dependent.

Dependent system A system of equations that has an infinite number of solutions. The graphs of the equations are all the same line.

Dependent variable A variable whose value is determined by the value of another variable in an equation.

Determinant The portion of the quadratic formula under the radical. Example: $b_2 - 4ac$

Diameter A line segment passing through the center of a circle and connecting two points on the circle.

Difference The answer to a subtraction operation.

Difference of two squares The binomial $a_2 - b_2$; it is a special product that factors into $(a + b)$ $(a - b)$.

Direct variation A situation in which one variable is proportional to a constant multiple of another. In the equation $y = 5x$, y varies directly with x.

Discriminant A number obtained from a matrix by calculating the difference of the products along the diagonals.

Disjunction A compound inequality involving the OR operator.

Distance formula For two points (x, y) and (a, b), the distance between them can be calculated by using the formula $(a - x)^2 + (b - y)^2$.

Distributive property The distributive property states that $a(b + c) = ab + ac$ for all values of a, b, and c. Example: $5(3 + 2) = 5 \cdot 3 + 5 \cdot 2$

Division property of equality When you divide both sides of an equation by the same nonzero quantity, the equation stays balanced.

Division property of inequality When you divide both sides of an inequality by the same positive number, the inequality stays balanced. When you divide both sides by the same negative number, you must flip over the inequality symbol, then the equation stays balanced. Division by zero is impossible.

Domain The set of all input values for a function; the allowable values of the independent variable.

Elements In a matrix or array, the individual numbers are called elements.

Entries In a table of information, the values within the table are called entries.

Equation A mathematical sentence stating that two quantities are equal.

Expanded notation A way of writing a number so that each digit is written separately and multiplied by a power of ten written in decimal form. All the digits and their multipliers are then written as one sum.

Experimental probability A probability based on the number of times an event has occurred over the number of trials that were performed.

Exponent A quantity representing the power to which a base is raised. Example: in 3^4, the exponent is 4; it means $3 \cdot 3 \cdot 3 \cdot 3$, or 81.

Exponential decay An exponential function in which either the exponent is less than one and the base is greater than one, or the exponent is greater than and the base is between zero and one.

Exponential function An equation in which at least one variable is used as an exponent.
Example: $y = 2^x$

Exponential growth An exponential function in which the exponent is greater than one and the base is greater than one.

Extraneous solution An extra solution obtained in the process of solving an equation. The extraneous solution is not a solution to the original equation.

Extremes In the proportion $\dfrac{a}{b} = \dfrac{c}{d}$, a and d are called the extremes.

Factor (noun) A part or piece of something. When parts are multiplied together to produce a quantity, each part is called a factor.

Factor (verb) To resolve a number or a polynomial into its individual factors.

Factor pair Two integers multiplied together to produce a number are considered a factor pair for that number.

Feasible solution Any point that makes all the inequalities in a linear programming problem true.

FOIL technique First, Outside, Inside, Last—a mnemonic for remembering how to multiply binomials.

Formula An equation that expresses a relationship between at least two variables.

Frequency distribution table A table showing the number of individual things in each of certain groups.

Function A relation between two variables such as x and y where there is only one possible value of y for any given value of x.

Function notation The notation used to represent the rule between two variables, usually written $f(x)$ and read "f of x."

Fundamental counting principle To find the number of ways to make a combination of choices, multiply together the number of possible choices in each category.

Greatest integer function A function that rounds input values down to the nearest integer.

Grouping symbols Parentheses or brackets used to set off one part of an equation from another.

Guess and check A method of problem solving in which a guess is made and checked in an equation. Based on the result, a new adjusted guess is made and then checked. This process is repeated until a suitable answer is found.

Histogram A bar graph made using the information from a frequency distribution table.

Horizontal line A line with slope zero. All horizontal lines are parallel to the x-axis.

Horizontal translation Moving the graph of a function to the right or left without changing the shape of the graph.

Hyperbola The graph of an inverse variation equation.

Hypotenuse In a right triangle, the side opposite the right angle; it is always the longest side.

Inconsistent system A system of equations that has no solutions.

Independent events Events whose outcomes have no effect on each other.

Independent variable A variable whose value does not depend on the value of any other variable.

Inequality A mathematical sentence stating that two quantities are greater than or less than one another.

Integer A number with no fractional part. The numbers $\ldots -2, -1, 0, 1, 2 \ldots$ represent the set of integers.

Intercept The point at which the graph of a line crosses an axis.

Interest Money charged for the use of money, usually calculated as a percentage of the amount of money borrowed.

Intersection The point at which two lines cross. Also, items that are common to two or more sets are said to be in the intersection of those sets. The AND operator indicates an intersection.

Interval A set of numbers between two given numbers. Example: the set of numbers between 3 and 8 is an interval

Inverse operations Operations that undo each other, such as addition and subtraction.

Inverse variation A situation in which one variable is proportional to the inverse of another. The equation $y = \dfrac{1}{x}$ shows y varying inversely with respect to x.

Irrational number A number whose decimal representation never terminates or repeats.

Isolating a variable The process of performing operations on an equation in order to leave a chosen variable on one side of the equation and have all other expressions on the other side.

Isosceles triangle A triangle that has two angles exactly the same.

Leg Either of the two sides of a right triangle adjacent to the right angle.

Like terms Terms that contain exactly the same variables raised to the same exponents. Example: $3x^2$ and $-2x^2$ are like terms

Line of best fit The line that comes closest to going through all the points on a graph.

Linear equations An equation written in the form $y = mx + b$, where m and b are constants. The graph of a linear equation is a straight line.

Linear function A function that can be represented by a straight line not parallel to the y-axis.

Linear programming A method of finding the best solution to a system of inequalities.

Linear term A polynomial, a term with a variable raised to the first power.

Magnitude The numerical value or length of a vector.

Matrix Another name for an array of numbers.

Mean Often called the average, it is the total of a number of quantities divided by the number of items.

Means In the proportion $\dfrac{a}{b} = \dfrac{c}{d}$, the means are b and c.

Measure of central tendency A number that somehow represents an entire set of numbers. The mean, median, and mode are all measures of central tendency.

Median The middle number of a set of data.

Metric System A system of weights and measures whose units are all based on powers of ten.

Mode The most common number in a set of data. If no number appears more than once, there is no mode.

Monomial A variable multiplied by a constant.

Multiplication property of equality When you multiply both sides of an equation by the same quantity, the equation remains balanced.

Multiplication property of inequality When you multiply both sides of an inequality by the same positive number, the inequality remains balanced. When you multiply by a negative number, then you must flip the inequality sign for the equation to stay balanced.

Multiplicative identity One is the multiplicative identity because when you multiply any quantity by one, the result is the same quantity. Example: $5 \cdot 1 = 5$

Multiplicative inverse Any two numbers whose product is one are multiplicative inverses; also called reciprocals. Example: 3 and $\frac{1}{3}$ are multiplicative inverses

Mutually exclusive events Two events where the occurrence of one event precludes the occurrence of the other.

Negative correlation A relationship where the increase of one variable implies a decrease in the other.

Negative exponent When a negative number is used as an exponent. Example: $x_{-n} = \frac{1}{x_n}$

Negative integer Any integer whose value is less than zero.

Numerical expression A mathematical expression that contains no variables.

Occurrence The outcome of one event.

Odds The ratio of the probability of an event's occurrence to the probability of its nonoccurrence.

Open interval An interval that does not include its endpoints.

Opposite of a vector The opposite of a vector A has the same magnitude as vector A but points in precisely the opposite direction.

Opposite of an integer Two integers are opposite if they are the same distance from zero. Example: 5 and -5 are opposites

Optimum solution In linear programming, the best of all the feasible solutions.

Or An operator that joins two mathematical sentences and indicates that solutions will satisfy *either* of the two sentences. The system would be a disjunction.

Order of operations A set of rules determining the way to evaluate a mathematical expression. First, perform all operations within any grouping symbols. Then, perform all calculations involving exponents. Multiply or divide in order from left to right. Finally, add or subtract in order from left to right.

Ordered pair A pair of numbers enclosed by parentheses and separated by a comma. The ordered pair represents a unique point in the Cartesian coordinate plane. Example: (2, 1) is an ordered pair; it is represented by the point 2 units to the right and 1 unit up from the origin

Origin The point in the Cartesian coordinate system where the axes cross and where the values of both x and y are zero.

Outcome The result of the occurrence of an event.

Outlier A data point that lies far away from all of the other data points in a set.

Parabola The graph of a quadratic equation.

Parallel lines Lines that have the same slope but different y-intercepts. Parallel lines never intersect.

Parent function The simplest form of a function that can be changed to generate an entire class of functions with similar graphs. Example: $y = x^2$ is the parent function of $y = x^2 + 2$

Pencil test Another name for the vertical-line test.

Percent Part of 100. Example: 60% is 60 out of 100; $60\% = 0.60 = 0.6$

Perfect square number A number that has one factor pair which is two of the same number.

Perfect square trinomial A quadratic of the form $(ax)_2 + 2abx + b_2$, where $a, b,$ and c are constants. A perfect square trinomial factors into a binomial square, $(ax + b)_2$.

Perpendicular Two lines are perpendicular if they meet at right angles to each other.

Pi The ratio of a circle's circumference to its diameter.

Polynomial A sum of two or more terms using one or more variables.

Positive correlation A relationship in which the increase of one variable implies an increase in the other.

Positive integer An integer greater than zero.

Power When you multiply the base b by itself n times, the product is the nth power of b. Example: the third power of 2 is 2^3, or 8

Power of a power When an exponential expression is raised to a power, the powers are multiplied together.

Power of a product When a product is raised to a power, each factor of the product is raised to the power.

Power of a quotient When a quotient is raised to a power, both the numerator and the denominator are raised to the power.

Power of ten If a number can be written in the form 10^n, where n is an integer, the number is a power of ten.

Power of ten notation A form where a quantity is written as a number multiplied by a power of ten.

Prime Anything that cannot be factored is called prime.

Prime factorization Breaking a number down into a list of factors that are all prime numbers.

Prime number A number whose only factor pair is one and the number itself.

Principal The basic amount of money used to calculate interest.

Probability The likelihood of an event's occurrence, often expressed as a ratio or a percent.

Product The result of multiplying two or more quantities.

Product of powers property When two or more exponential expressions involving the same base are multiplied together, the result is the base written with the sum of the exponents as its power.

Property of adding zero When you add zero to any quantity, the result is the same quantity. Example: $5 + 0 = 5$

Property of multiplying by one When you multiply any quantity by one, the result is the same quantity. Example: $5 \cdot 1 = 5$

Proportion A mathematical sentence stating that one ratio is equal to another.

Example: $\frac{a}{b} = \frac{c}{d}$, where $b \neq 0$ and $d \neq 0$

Proportion property In a true proportion, the product of the means equals the product of the extremes.

Pythagorean Theorem In a right triangle, the square of the length of the hypotenuse equals the sum of the squares of the lengths of the two legs.

Pythagorean triple A set of three *integers* that satisfies the Pythagorean Theorem.

Quadrants The Cartesian coordinate plane is divided into four regions by the axes; these regions are called quadrants.

Quadratic equation An equation in which the highest power of the independent variable is two.

Quadratic formula A formula used to calculate the roots of a quadratic equation:

$$x = \frac{-b \pm \sqrt{b^2 - 4ac}}{2a}$$

Quadratic function A quadratic equation expressed as a function. The parent function is $y = x^2$.

Quadratic term In a polynomial, the term whose variable is raised to the second power.

Quotient The result of a division operation.

Quotient of powers property When an exponential expression is divided by another expression involving the same base, the result is the base raised to the difference of the exponents.

Radical A sign ($\sqrt{}$) indicating the square root of a quantity.

Radicand The quantity under the square root sign.

Radius A line segment from the center of a circle to the circle itself.

Range (of data) The interval in which all the data lie, found by calculating the difference between the data points with the largest and the smallest values.

Range (of function) The set of all outputs for a function; the possible values of the dependent variable.

Rate of change The slope of a linear equation.

Rate of interest The ratio of the interest charged to the principal borrowed expressed as a percent.

Ratio The relative sizes of two numbers, usually expressed as $a : b$ or $\frac{a}{b}$, where $b \neq 0$.

Rational expression An expression that is a ratio of two quantities.

Rational number Any number that can be written as a ratio of two integers with the second integer not equal to zero. Integers, terminating decimals, and repeating decimals are all rational numbers.

Rationalizing Removing radicals from the denominator of a fraction by multiplying both the numerator and the denominator by the radical.

Real numbers The set of all rational and irrational numbers combined is called the set of real numbers.

Reciprocal If the product of two numbers is one, the numbers are reciprocals. A number's reciprocal is also called its multiplicative inverse.

Reciprocal function The function $y = \frac{1}{x}$, where $x \neq 0$.

Regression line Another name for the line of best fit.

Relation Any set of ordered pairs.

Repeating decimal A decimal that after some point consists of an infinite number of repetitions of the same number sequence.

Example: $2\frac{2}{3} = 2.666\ldots$

Restrictions Limitations imposed on the domain and range of a function so that all the possible inputs and outputs make sense for a real-world situation.

Resultant The one vector that represents the combination of two or more vectors.

Right angle An angle measuring exactly 90 degrees, sometimes called a square angle.

Right triangle Any triangle containing a right angle.

Rise of a line The vertical change in a line between two points, found by calculating the difference between the y values of the points.

Roots The values of x that make a polynomial equal zero, also called the zeros of a polynomial.

Run of a line The horizontal change in a line between two points, found by calculating the difference between the x values of the points.

Sample space The set of all possible outcomes for a given event.

Scale factor The ratio by which the original dimensions of a figure are multiplied to enlarge or reduce the figure.

Scatter plot A graph treating data as individual points, used to check for correlation in data.

Scientific notation A shorthand way of writing very large or small numbers such that they are represented by a number between one and ten multiplied by a power of ten written in exponential form. Example: $3,700,000 = 3.7 \times 10^6$

Similar triangles Two triangles that have the exact same angles are called similar triangles.

Simple interest Interest calculated only on the original principal.

Sine The ratio of the opposite side of an angle to the hypotenuse in a right triangle.

Slope The ratio of the rise of a line to its run; slope can be thought of as the "steepness" of the line.

Slope-intercept form The graph of a linear equation written in the form $y = mx + b$ has m as its slope and b as its y-intercept.

Special product Perfect square trinomials and differences of squares are called special products.

Special right triangles Right triangles with known relationships between the sides and the angles.

Square of a number A number raised to the second power. Example: 5^2 is five squared

Square root If $y = x^2$, then x is the square root of y. The square root of a is written $\sqrt{a}$. Example: the square root of 25, $\sqrt{25}$, is 5

Square root property The solution for $x^2 = k$, where $k > 0$ is $\sqrt{k}$ or $-\sqrt{k}$.

Standard form For a linear equation, $ax + by = c$. For a quadratic equation, $ax_2 + bx + c = d$.

Statistics The study of interrelationships among data.

Substitute Replacing a variable with any quantity that has the same value.

Subtraction property of equality If the same quantity is subtracted from both sides of an equation, then the equation stays balanced.

Subtraction property of inequality If the same quantity is subtracted from both sides of an inequality, the inequality stays balanced.

Sum (Σ) The result of adding two or more quantities.

System of equations Two or more interrelated equations involving the same variables.

System of inequalities Two or more inequalities involving the same variables and usually related by AND or OR.

Tangent The ratio of the opposite side of an angle to the adjacent side in a right triangle.

Terminating decimal A decimal that stops after a certain number of digits. Example: $\frac{3}{4} = 0.75$

Test point A value substituted into an inequality to find out on which side of a line the solutions lie.

Theoretical probability The ratio of the number of favorable outcomes to the number of outcomes possible.

Transformation Changing the graph of a parent function by multiplying it by a constant or by adding or subtracting a constant.

Translation Moving a graph or figure up or down or to the left or right without altering its slope.

Tree diagram A diagram showing the successive decomposing of a number into its factors.

Trigonometric ratios The ratios of the lengths of the sides in a right triangle. Example: the tangent is the ratio of the opposite side to the adjacent side

Trigonometry The study of the relationships among the angles and the sides in triangles.

Trinomial A polynomial with three terms.

Union The union of two sets is the set of elements that are contained in either of the sets. A union is indicated by the OR operator.

Variable A quantity, usually represented by a letter, that can take on different values.

Vectors A quantity having both magnitude and direction, often represented by an arrow.

Venn diagram A diagram that uses circles to represent sets and their interrelations.

Vertex The point at which a parabola reverses direction; also the "tip" of the absolute value function's graph.

Vertex (of a triangle) Another name for the corners of a triangle.

Vertical line A line with undefined slope. All vertical lines are parallel to the y-axes.

Vertical-line test If *any* vertical line crosses a graph in more than one place, then the graph is not a function.

Vertical reflection An image in which each point of a graph is replaced with a point an equal distance on the other side of the x-axis. It looks as if the original graph has been turned upside-down.

Vertical translation Moving a graph up or down without changing the shape of the graph.

Volume The amount of space inside a three-dimensional object such as a box or a sphere.

X-axis The horizontal axis of the Cartesian coordinate system.

X-coordinate The first number in an ordered pair. It tells how far to the right or left of the origin a point is located.

X-intercept The *x*-coordinate of the point at which a line crosses the *x*-axis.

Y-axis The vertical axis of the Cartesian coordinate system.

Y-coordinate The second number in an ordered pair. It tells how far up or down from the origin a point is located.

Y-intercept The *y*-coordinate of the point at which a line crosses the *y*-axis.

Zero exponent Any quantity (except zero) raised to the zero power always equals one.

Zero power The zero power of any number equals one. Example: $3^0 = 1$

Zero product property If two quantities' product is zero, then one or both of the quantities must be zero.

Zeros The values for *x* that make a polynomial equal zero, also called the roots of the polynomial.

Selected Answers

This section contains selected answers to the Practice and Problem Solving and Mixed Review sections.

Chapter 7 Statistics and Probability

Lesson 7.1, pages 4-10

7. 10   **9.** 70   **11.** 75.8   **13.** 2.4
15. 262   **17.** 276.3
19. 60 kilograms   **21.** 59.4 kilograms
23. It will not affect the mode—60 still occurs the most. It will not affect the median—60 is still the middle term.
25. $5200   **27.** $1127.45

Lesson 7.2, pages 11-17

5.

Class	Tally	Frequency
0 - 9	////	4
10 - 19	//////	6
20 - 29	///////	7
30 - 39	///	3
	$n = 20$	

7. 31   **9.** 20

11.

Classes	Tally	Frequency	$(c)(f)$
50	//	2	100
60	/	1	60
70	//	2	140
75	///	3	225
80	//////	6	480
85	/////	5	425
90	/	1	90
95	///	3	285
100	//	2	200
		25	2005

13.

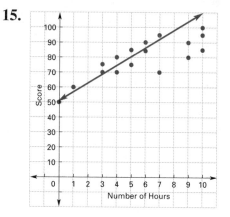

23. $105   **25.** 40

27.

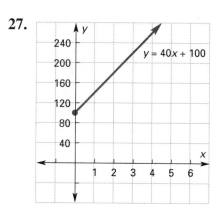

Lesson 7.3, pages 18-23

5. The data shows a negative correlation.

7. One example is $y = -\dfrac{3}{5}x + 26$

9. The answer depends on the line drawn.

11. 80   **13.** 2 students; 9 hours and 6 hours

15.

17. $y = 2.5x + 65$

19.

January	−$45.00
February	−$75.00
March	+$40.00
April	+$ 5.00
May	+$20.00

21. about 216.8 square feet   **23.** 19.5

Lesson 7.4, pages 24-30

5. 25% **7.** 55% **9.** 88%
11. about 41.8% **13.** 9.1% **15.** The answer
depends on the data supplied. **17.** 50%
19. (h, h, h), (h, h, t), (h, t, t), (t, t, t), (t, t, h),
(t, h, h), (t, h, t), (h, t, h)
21. 3 to 5, or 60% **23.** 80°
25. -20, -19, and -18 **27.** (2, 0)

Lesson 7.5, pages 31-38

3. Answers depend on individual results.
5. about 50 each **7.** (T, H) represents tails for
   the nickel and heads for the penny.
9. (T, H), (H, T), (H, H), (T, T)
11-19. Answers depend on individual results.
21. 24 miles **23.** 21 defective parts
25. $y = 25,000 - 750x$, where x represents the
   number of seconds the plane descends and
   y represents the height at any time x.

Lesson 7.6, pages 39-45

5. $\dfrac{1}{36}$ **7.** $\dfrac{1}{4}$ **9.** $\dfrac{11}{36}$ **11.** $\dfrac{1}{36}$ **13.** $\dfrac{15}{36}$
15. $\dfrac{1}{6}$ **17.** $\dfrac{5}{6}$ **19.** $\dfrac{36}{55}$ **21.** $\dfrac{17}{20}$

23. $r = \dfrac{i}{pt}$ **25.** In April 800 phones were
made, in May 1000 phones were made, and in
June 2000 phones were made.

Lesson 7.7, pages 46-51

 3. 24 **5.** 12 **7.** 10 **9.** 60 **11.** 45
13. 30 **15.** 210 **17.** $1150 **19.** $50
21.

23.

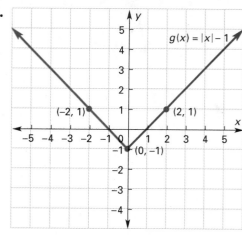

Lesson 7.8, pages 52-57

3. $\dfrac{1}{8}$ **5.** $\dfrac{17}{24}$ **7.** $\dfrac{19}{24}$ **9.** $\dfrac{3}{8}$ **11.** $\dfrac{1}{8}$
13. $\dfrac{5}{33}$ **15.** $\dfrac{25}{144}$ **17.** $\dfrac{1}{16}$ **19.** $\dfrac{9}{16}$
21. 0 **23.** $\dfrac{1}{22}$ **25.** $\dfrac{1}{11}$ **27.** $\dfrac{1}{12}$
29. 11 lamps **31.** $y = -2400x + 26,000$;
The car is 4 years old.

Chapter 8 Systems of Equations

Lesson 8.1, pages 4-10

5.

7.

13.

9.

15.

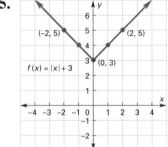

17. $y \geq 3$ **19.** 4, 5, 6, 6, 6, 7, 7, 7, 8, 8
21. 6.5

Lesson 8.2, pages 11-17

7. consistent and independent
9. consistent and dependent
11. inconsistent   **13.** consistent and
independent   **15.** inconsistent
17. inconsistent   **19.** independent
21. A. City messenger   +2.0 miles
    B. Dentist   −1.2 miles
    C. Hospital nurse   +3.3 miles
    D. Hotel employee   +1.2 miles
    E. Secretary   +0.2 miles
    F. Teacher   −0.3 miles
23. about 4.002389×10^7 kilometers
25. about 5.099×10^{14} square kilometers
27. The circumference is increased by a factor of
2. The surface area is increased by a factor of 4.

11.

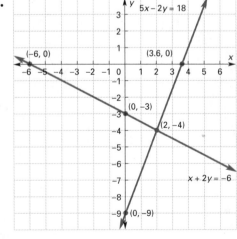

Lesson 8.3, pages 18-24

5. dependent   **7.** $(11, 1)$ **9.** $(8, 0)$
11. $(3, -2)$ **13.** The 3-subject notebook cost is \$5.00. The single-subject notebook cost is \$2.50.   **15.** Joe does 37 haircuts per week, and Nita does 58 haircuts per week.
17. $360°$ **19.** $45°, 65°,$ and $70°$ **21.** $\frac{1}{2}$
23. $\frac{7}{10}$ **25.** $\frac{1}{4}$ **27.** $\frac{9}{100}$

Lesson 8.4, pages 25-31

5. $(0, 0)$ **7.** $(3, 3)$ **9.** $(2, 0.5)$ **11.** $(3, 1)$
13. One delivery will contain 13 pallets, the other will contain 10 pallets.
15. The flat rate for the first 10 words is \$10, each additional word is \$0.65.
17. One truck will hold 2.87 tons of sand.
19. $N = \dfrac{x}{(215)(2.87)}$ **21.** 2   **23.** $0.5 \le x \le 5$
25. As x becomes very small, $g(x)$ becomes very large.

Lesson 8.5, pages 32-37

5. $(-11, -6)$ **7.** $(-1, -2)$ **9.** $(10, 14)$
11. $(0, -2)$ **13.** dependent   **15.** The sprayer costs \$5 per hour to rent. The generator costs \$10 per hour to rent.   **17.** One compartment holds 27 pieces of baggage, and the other compartment holds 68 pieces of baggage.
19. 9.5×10^{-8} centimeters   **21.** 3.0×10^{10} centimeters per second   **23.** 1.5 square feet
25. One machine part weighs 1.8 pounds, the other part weighs 6.8 pounds.

Lesson 8.6, pages 38-43

3. $\left(-\frac{2}{3}, -\frac{5}{3}\right)$ **5.** $(0, 4)$ **7.** $\left(\frac{5}{3}, \frac{5}{3}\right)$ **9.** $(-3, 2)$
11. $\left(\frac{2}{3}, -3\right)$ **13.** The speed of the truck is 50 miles per hour. The speed of the car is 65 miles per hour.   **15.** 3   **17.** 7.6   **19.** 130.5 rentals

---

21. B = Boy; G = Girl sample consists of
(B, B, B, B), (B, B, B, G), (B, B, G, G),
(B, G, G, G), (G, G, G, G), (G, G, G, B),
(G, G, B, B), (G, B, B, B), (G, B, B, G),
(G, B, G, B), (B, G, B, G), (B, G, G, B),
(G, G, B, G), (G, B, G, G), (B, G, B, B),
(B, B, G, B)
23. $\frac{3}{8}$ **25.** $\frac{7}{8}$

Chapter 9 Inequalities

Lesson 9.1, pages 4-8

5. $x < 5$ **7.** $x \le 0$ **9.** $2 < x \le 7$
11.

13.
15.
17. $x < 25.5$ **19.** $1\frac{1}{2} < x < 2$
21. 1.24×10^6 Newtons per square meter
23.
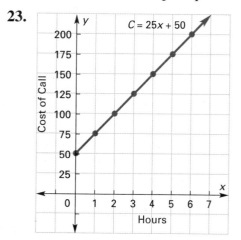

Lesson 9.2, pages 9-13

5.

$x > -6$
7.
$x > -5$

9.

$x < 6$

11.

$x \geq 1$

13. $x + 35 \geq 150$; Tai needs to save at least $115 more to purchase the stereo.

15. $x - 15 < 50$; The larger number is any number less than 65.

17. $x + 210 \leq 350$; The gift shop can spend $140 or less in advertising.

19. 44   **21.** 41   **23.** $\frac{1}{23} \approx 4.3\%$

25. about 0.01%

Lesson 9.3, pages 14-19

3.

5.

7.

9.

11.

13. $x > 30\% \, (75)$; at least $22.50

15. $8x \leq 1000$; $125   **17.** $86 + 88 + x \geq 270$; 96 points   **19.** $20 - 1.39x \leq 5$; less than or equal to 10.8 gallons   **21.** 3

23. $y = 3x + 11$ **25.** 0

Lesson 9.4, pages 20-26

5.

7.

9.

11.

13. $x > 72$; above $72°$

15. $145 < 2x + 5 < 155$; between $70 and $75

17. 152.4 pounds   **19.** positive correlation

21. $(\frac{1}{2}, 2)$ **23.** $x = y + 20, x + y = 100$; Mario saved $40 and Nancy saved $60.

25. $15x + 12y = 34.20, 10x + 7y = 21.45$; One gallon of 87 octane gas costs $1.20, and one gallon of 92 octane gas costs $1.35.

Lesson 9.5, pages 27-30

5.

7.

9.

11.

13. $|x - 120| > 5$; over 125 minutes or under 115 minutes   **15.** $|x - 30| > 3$; less than 27 psi and greater than 33 psi   **17.** 8.75×10^6 cubic feet   **19.** 875 fans

21. $9 - 0.25(9) = x$; $6.75   **23.** $x + (x + 1) + (x + 2) = 63$; 20, 21, and 22

Lesson 9.6, pages 31-39

5.

7.

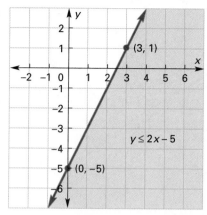

$y \leq 2x - 5$

9.

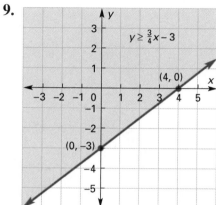

$y \geq \frac{3}{4}x - 3$

11.

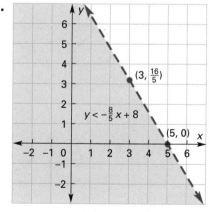

$y < -\frac{8}{5}x + 8$

13. $2L + 2W > 24$ **15.** 10 feet, 8 feet; 11 feet, 9 feet; 12 feet, 10 feet

17.

$y < -x + 30$

$$y < -x + 30$$

19. $10x + 6y \leq 60$ **21.** 0 jackets on the faster machine, 10 jackets on the slower machine. Two more solutions that satisfy the constraints are (3, 2) and (4, 3). **23.** The slope of line AB is -1. The slope of line MN is 1.

25. $y = x + 3$

27.

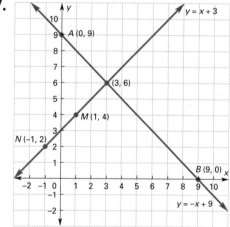

29. $y = -x$

Lesson 9.7, pages 40-44

3.

5.

11.

7.

13.

9.

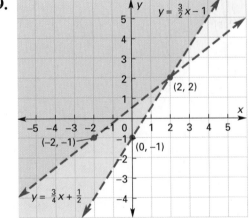

15. $x + y \leq 14;$ $(0.50)x + (1)y < 10$

17. To maximize profits, the corner point $(8, 6)$ should be used. The profit at this point will be $3(8) + 4(6) = \$48$.

19. Between 0 and 66 tennis shoes can be produced, and between 0 and 100 running shoes can be produced. Any combination in the solution will work.

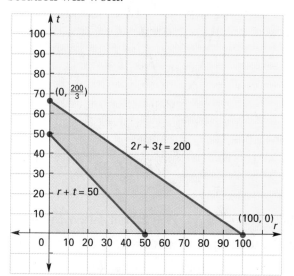

21. $\dfrac{3}{10} = 30\%$ **23.** $\dfrac{9}{245} \approx$ about 3.67%
25. $\dfrac{19}{50} = 38\%$ **27.** Tim has invested \$42. Shaneka has invested \$54.

Chapter 10 Polynomials and Factors

Lesson 10.1, pages 4-9

5. binomial **7.** binomial **9.** trinomial **11.** $5x + 8$ **13.** $3x^2 - x - 3$
15. $x^2 + 5x - 12$ **17.** $4x^2$ **19.** $x + 3$
21. $x^2 + 7x + 4$ **23.** $4x^2 - 8x$
25. $6x^2 - 2x + 7$ **27.** $r = \dfrac{d}{t}$
29. 9 **31.** -2 **33.** 3 CDs

Lesson 10.2, pages 10-15

7. $1 \cdot 45; 3 \cdot 15; 5 \cdot 9$ **9.** $1 \cdot 32; 2 \cdot 16; 4 \cdot 8$
11. $1 \cdot 26; 2 \cdot 13$ **13.** $2 \cdot 3 \cdot 7$ **15.** $2 \cdot 3 \cdot 3$
17. $2 \cdot 2 \cdot 7$ **19.** 6 **21.** 12 **23.** 13 **25.** 6
27. $(-1, -3)$ **29.** The pieces are 32.5 inches and 35.5 inches.

Lesson 10.3, pages 16-20

3. $ab + 3a$ **5.** $7x^2 + 35xy$
7. $8x^4 + 14x^2 - 10x$ **9.** $x^2 + 2y^2$
11. $x^3y - 2xy^3$ **13.** $V = 3\pi r^3$ **15.** $y(3x - 5y)$
17. $cx^2(x + 2x^2)$ **19.** $\dfrac{13}{20}$
21. all whole numbers less than 7
23. all whole numbers less than 4
25. $\dfrac{3}{500} = \dfrac{12}{n}$; \$2000

Lesson 10.4, pages 21-26

5. a^7 **7.** $10x^7$ **9.** $\dfrac{8x^3}{27}$ **11.** $6x^5 + 8x^3$
13. $4s$ **15.** $C = \dfrac{Dx}{48}$
17. $2w + 2(w + 4) = 36$; 77 square inches
19.

Lesson 10.5, pages 27-31

5. $\dfrac{5x}{4}$ **7.** $\dfrac{15 - 8x}{10x}$ **9.** $\dfrac{x^2 + 2}{2x}$ **11.** $\dfrac{-x + 8}{x^2}$
13. $\dfrac{3x^2 - x + 4}{2x^2}$ **15.** 10 hours
17. 75 hours **19.** 24, 38, and 29 students
21. The shirts are \$35, and the shorts are \$22.

Lesson 10.6, pages 32-38

5. $x^2 + 7x + 12$ **7.** $2x^2 + 5x + 2$
9. $x^2 + 7x + 10$ **11.** $3x^2 + 10x + 3$
13. $10x^2 - 13x - 3$ **15.** $15x^2 - 11x + 2$
17. -9 **19.** \$50 **21.** $2L + 2W = 19$ and $4W - 3 = L$; The width is 2.5 centimeters. The length is 7 centimeters.

Lesson 10.7, pages 39-41

3. $(a + 1)(a + 7)$ **5.** $(x - 1)(x - 1)$
7. $(a + 3)(a - 2)$ **9.** $(w - 2)(w - 4)$
11. $(3x + 5)(2x - 3)$ **13.** 18 square feet
15. ± 5 **17.** -12 **19.** $\dfrac{1}{273}$

Lesson 10.8, pages 42-46

5. $(x + 2)(x + 3)$ **7.** $(x - 2)(x - 3)$
9. $(x - 1)(x + 6)$ **11.** $(x + 1)(5x + 2)$
13. $(3x + 5)(x + 3)$ **15.** $(2x + 5)(3x - 4)$
17. $x + 15 = 18; \$3$ **19.** $25x + 10x = 280;$
8 feet   **21.** $(-4, 1)$

Lesson 10.9, pages 47-52

7. $x^2 - 9$ **9.** $x^2 - y^2$ **11.** $x^2 - 9$
13. $4y^2 + 12xy + 9$ **15.** $(x + 5)(x - 5)$
17. $(3 - 4y)(3 + 4y)$ **19.** $(y - 9)^2$
21. $(y - 1)^2$ **23.** $(x - a)^2$ **25.** $b = \dfrac{P - 2a}{2}$
27. $y = \dfrac{1}{5}x + \dfrac{3}{5}$ **29.** $y = -3$ **31.** $(-\dfrac{1}{2}, -2)$

Chapter 11 Quadratic Functions
Lesson 11.1, pages 4-9

7. $y = 3x^2 + 5$, translation 5 units up, shrink by
a factor of 3, vertex $(0, 5)$

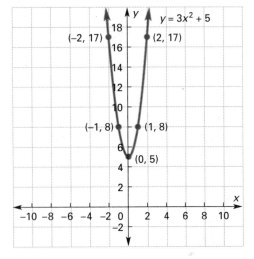

9. $y = (x - 2)^2 + 1$, translation 2 units right
and 1 unit up, vertex $(2, 1)$

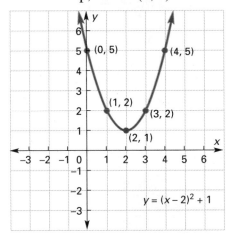

11. $y = -\dfrac{1}{2}(x + 1)^2 + 2$, translation 1 unit left
and 2 units up, reflection across the x-axis,
shrink by a factor of $\dfrac{1}{2}$, vertex $(-1, 2)$

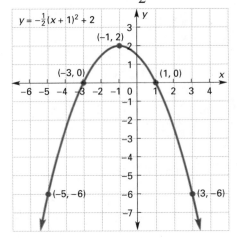

13. $\$78.41$ **15.** $x = y + 2$ and $4x + 4y = 40;$
4 inches and 6 inches
17. $(85 + 90 + 93 + x) \div 4 \geq 90; 92$
19. $y = \dfrac{3}{2}x + \dfrac{5}{2}$

Lesson 11.2, pages 10-16

7. $-2; -4; V(-3, -1)$ **9.** $-1; V(-1,1)$
11. $2; 5; V(-\dfrac{3}{2}, -\dfrac{13}{4})$ **13.** $5; -1; V(2, -9)$
15. $1; 3; V(2, -1)$ **17.** $4.98; 2.56;$
$V(1.7, -4.37)$ **19.** $\$900$ **21.** 420 cubic
centimeters   **23.** 2   **25.** $-1; 6$ **27.** $(0, 6)$

Lesson 11.3, pages 17-22

3. $-2; -3$ **5.** $-3; 1$ **7.** $-2; -1$ **9.** $\frac{3}{2}; -4$

11. $-8; 2$ **13.** $-\frac{3}{2}; \frac{4}{3}$

15. $5^2 = 4s + 12$; 6 meters   **17.** 6×10^{25}

19. $2(W + 1) + 2W = 180$

21. $2L + 2W = 180; L - W = 1$

23. the same   **25.** 0

Lesson 11.4, pages 23-29

5. $y = -(x - 4)^2 + 16$; maximum = 16;
$V(4, 16); x = 4$

7. $y = -(x - \frac{3}{2})^2 + \frac{17}{4}$; maximum $= \frac{17}{4}$;
$V(\frac{3}{2}, \frac{17}{4}); x = \frac{3}{2}$

9. $y = -(x + \frac{7}{2})^2 + \frac{89}{4}$; maximum $= \frac{89}{4}$;
$V(\frac{7}{2}, \frac{89}{4}); x = -\frac{7}{2}$

11. $y = -(x + 2)^2 - 8$; maximum $= -8$;
$V(-2, -8); x = -2$ **13.** 8%

15. $4x + 2y = 2880$ and $x + y = 960$;
480 hammers

Lesson 11.5, pages 30-33

5. $-2; 6$ **7.** $-1.32; 5.32$ **9.** $0.8; -8.8$

11. $-1.3; 2.3$ **13.** $-0.54; 5.54$

15. $-4.24; 0.24$ **17.** $0; 8$ **19.** 10 meters by
12 meters   **21.** \$1,050,000   **23.** $\frac{1}{8}$ **25.** $\frac{1}{2}$

Lesson 11.6, pages 34-41

3. 13   **5.** 0   **7.** $-3; 7$ **9.** none   **11.** -1.1;
0.22   **13.** none   **15.** 7 meters by 12 meters
17. 24 inches and 36 inches   **19.** 328 sensors
21. 7.75 seconds   **23.** 11   **25.** 18
27. 19

Chapter 12 Right Triangle Relationships
Lesson 12.1, pages 4-11

3. 1; 12   **5.** $17\frac{1}{3}; 39$ **7.** 144 yards
9. 12.5 inches; 18.75 inches   **11.** 10 feet
13. undefined   **15.** $y = -2$

17.

$x \le 3$

-6	-5	-4	-3	-2	-1	0	1	2	3	4	5	6	7	

19. $10y - y > 16$; at least 17.7 feet

Lesson 12.2, pages 12-17

3. no   **5.** yes   **7.** no   **9.** 21.9 yards
11. 50.3 feet   **13.** 5   **15.** 17
17. $(AB)^2 + (AC)^2 = (BC)^2$; $160 = 160$
19. \$216   **21.** 2 gallons; \$9

Lesson 12.3, pages 18-22

3. $2\sqrt{3}$ **5.** $10\sqrt{2}$ **7.** $\frac{\sqrt{6}}{3}$ **9.** $3\sqrt{2}$

11. $r = \sqrt{\dfrac{A}{\pi}}$ **13.** $20\sqrt{15}$ feet per second

15.+17.

$y = \sqrt{x}$		$y = \sqrt{2x}$		$y = \sqrt{\frac{1}{2}x}$	
x	y	x	y	x	y
0	0	0	0	0	0
1	1	1	$\sqrt{2}$	1	$\frac{\sqrt{2}}{2}$
4	2	4	$2\sqrt{2}$	4	$\sqrt{2}$
9	3	9	$3\sqrt{2}$	9	$\frac{3}{2}\sqrt{2}$
16	4	16	$4\sqrt{2}$	16	$2\sqrt{2}$

19. $2L + 2(L - 25) = 250$; 75 by 50 feet
21. $x + y = 24$ and $y = 2x + 6$; 18 feet; 6 feet

Lesson 12.4, pages 23-32

5. $a, c,$ and d **7.** $10\sqrt{2}$ feet   **9.** $8\sqrt{2}$ inches
11. 12 feet; $6\sqrt{3}$ feet   **13.** 10 inches; $10\sqrt{3}$
inches   **15.** 180 inches   **17.** about 127 feet
19. 4 trucks   **21.** 1200   **23.** $x + y = 30$ and
$y = x + 6$; 18 bolts
25. $n^2 + (n + 2)^2 = (n + 1) + 155$; 8 and 10

Lesson 12.5, pages 33-44

5. $\sin A = \frac{4}{5}$, $\sin B = \frac{3}{5}$; $\cos A = \frac{3}{5}$, $\cos B \frac{4}{5}$;
$\tan A = \frac{4}{3}$, $\tan B = \frac{3}{4}$ **7.** $\sin A = \frac{\sqrt{2}}{2}$,
$\sin B = \frac{\sqrt{2}}{2}$; $\cos A = \frac{\sqrt{2}}{2}$, $\cos B = \frac{\sqrt{2}}{2}$;
$\tan A = 1$, $\tan B = 1$ **9.** 0.707

11. $\sin 35° = \frac{40}{x}$; about 69.74 feet

13. about 47.67 feet

15. $\tan 87° = \frac{x}{55}$; about 1049 feet

17. $-\frac{11}{2}$

19.

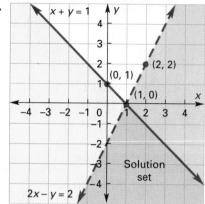

21. −4

Index

Credits

Chapter 7

2: Archive Photos. 3: Rhoda Sidney/PhotoEdit. 4: Lawrence Migdale/Stock Boston. 5: Alan Brown/Photonics Graphics. 9: Hewlett-Packard. 11: Aaron Chang/Stock Market. 15: David Parker/Photo Researchers, Inc. 17: Coco McCoy/Rainbow. 18: Matthew McVay/Stock Boston. 20: Michael Newman/PhotoEdit. 23: David M. Grossman/Photo Researchers, Inc. 24: George Goodwin/Picture Cube. 28: Loren Santow/Tony Stone Images. 31: Hill/The Image Works. 38: © 1995 PhotoDisc, Inc. 41: Daemmrich/The Image Works. 42: © 1995 PhotoDisc, Inc. 44: COREL. 46: Nancy Sheehan/PhotoEdit. 51: Mary Kate Denny/PhotoEdit. 57: Ford Aerospace and Communications Corporation. 61: CORD. 64: David R. Frazier/Photo Researchers, Inc. 67: Nathan Benn/Stock Boston. 69: Jeff Greenberg/Photo Researchers, Inc. 72: © 1995 PhotoDisc, Inc. 73: Cathy Ferris. 75: Charlie Westerman/Liaison International. 76: Bachman/Stock Boston.

Chapter 8

2: COREL. 3: © 1995 PhotoDisc, Inc. 4: Jon Riley/Tony Stone Images. 7: © 1995 PhotoDisc, Inc. 9: Richard Clintsman/Tony Stone Images. 9: © 1995 PhotoDisc, Inc. 12: David Young Wolff/PhotoEdit. 16: Courtesy of Delta Airlines. 18: John Coletti/The Picture Cube. 22: Tommaso Guicciardini/Photo Researchers, Inc. 23: Miro Vintoniv/The Picture Cube. 25: Jim Pickerell/Tony Stone Images. 28: Lien/Nibauer/Liaison International. 30: © 1995 PhotoDisc, Inc. 31: COREL. 33: Bob Daemmrich/Stock Boston. 36: Jeff Greenberg/Photo Researchers, Inc. 38: Andrew Errington/Tony Stone Images. 41: Tom Pantages. 42: © 1995 PhotoDisc. 44: CORD. 45: CORD. 49: © 1995 PhotoDisc, Inc. 51: Mark Burnett/Photo Researchers, Inc. 53: © 1995 PhotoDisc Inc. 55: top, David Shopper/Stock Boston; bottom, © 1995 PhotoDisc, Inc. 58: Tony Stone Images. 60: Stanley Rowin/The Picture Cube.

Chapter 9

2: Eurotunnel/QA Photos 3: Eurotunnel/QA Photos. 6: William Taufic/The Stock Market. 8: top, © 1995 PhotoDisc, Inc.; bottom, Stephen McBrady/PhotoEdit. 12: center, Nik Kleinberg/Stock Boston. 14: John Henley/The Stock Market. 16: Frank Siteman/Stock Boston. 17: Jeff Greenberg/PhotoEdit. 18: Richard Pasley/Stock Boston. 20: Jeff Persons/Stock Boston. 23: Dennis MacDonald/PhotoEdit. 25: Michele Bridwell/PhotoEdit. 27: © 1995 PhotoDisc, Inc. 29: © 1995 PhotoDisc, Inc. 31: © 1995 PhotoDisc, Inc. 35: Mark Gibson. 39: The Image Works. 40: Tom Pantages. 41: D. Chidester/The Image Works. 43: Henry Horenstein/Stock Boston. 52: David Carriere/Tony Stone Images. 54: Ed Lallo/Liaison International. 57: top, Seth Resnick/Liaison International; bottom, Mark Segal/Tony Stone Images. 59: Stuart McCall/Tony Stone Images. 60: Tony Stone Images. 62: Scott Slobodian/Tony Stone Images.

Chapter 10

2: © 1995 PhotoDisc, Inc. 3: © 1995 PhotoDisc, Inc. 4: Tom Carroll. 7: Phyllis Picardi/Stock Boston. 9: Granitsas/The Image Works. 15: Phil Degginger/Tony Stone Images. 17: Courtesy of Drexel University. 20: Eric Brissaud/Liaison International. 23: Tom Pantages. 25: Charles Gupton/Stock Boston. 27: Robert Reichert/Liaison International. 31: bottom, Pedrick/The Image Works. 32: David Young-Wolff/PhotoEdit. 38: Lee Snider/The Image Works. 39: CORD. 45: Robert Ginn/The Picture Cube. 46: Mark Gibson. 52: Matthew Borkoski/Stock Boston. 55: CORD. 58: Hewlett-Packard. 59: Jane Lotta/Photo Researchers, Inc. 61: Robert Fried/Stock Boston. 62: Miro Vintoniv/The Picture Cube.

Chapter 11

2: Tom Carroll. 3: Liaison International. 5: Bob Daemmrich/Stock Boston. 9: Reinstein/The Image Works. 10: Tom Carroll. 16: William Johnson/Stock Boston. 19: Esbin-Anderson/The Image Works. 21: Cary Wolinsky/Stock Boston. 23: Gregg Grosse. 27: Bob Daemmrich/The Image Works. 29: Sheila Nardulli/Liaison International. 31: Addison Geary/Stock Boston. 32: Courtesy of Tennessee Photo Services. 33: © 1995 PhotoDisc, Inc. 36: © 1995 PhotoDisc, Inc. 39: © 1995 PhotoDisc, Inc. 40: © TRW Inc. 42: CORD. 48: CORD. 50: © 1995 PhotoDisc, Inc. 52: top, Lynn Stone/Natural Selection; bottom, Giboux/Liaison International. 53: Oil Spill Public Information Center. 55: William Taufic/The Stock Market. 58: James Mejuto. 63: © 1995 PhotoDisc, Inc.

Chapter 12

2: COREL. 3: COREL. 6: COREL. 8: Blair Seitz/Photo Researchers, Inc. 11: Navy photo by Gerry Winey. 15: COREL. 16: Dick Davis/Photo Researchers, Inc. 17: Thomas Braise/The Stock Market. 18: Dennis MacDonald/PhotoEdit. 21: © 1995 PhotoDisc, Inc. 25: Tony Arruza/The Image Works. 29: Robert Rathe/Stock Boston. 31: © 1995 PhotoDisc, Inc. 33: © 1995 PhotoDisc, Inc. 38: © 1995 PhotoDisc, Inc. 40: © 1995 PhotoDisc, Inc. 42: Collins/The Image Works. 48: CORD. 49: CORD. 52: © 1995 PhotoDisc, Inc. 54: © 1995 PhotoDisc, Inc. 55: © 1995 PhotoDisc, Inc. 60: COREL. 62: Mark Gibson. 64: Tim Lynch/Stock Boston.